PLUMBING

ALSO IN THE REVISION AND SELF-ASSESSMENT SERIES

LEEDS COLLEGE OF BUILDING

Ideal for the student working alone, these new books allow readers to test their understanding of key subjects. Each topic begins with a summary of key facts and figures and follows with multiple-choice assessments. Students find these books stimulating and useful revision aids.

- Contains reference notes and definitions
- Includes multiple-choice questions
- Allows students to assess their progress
- Contains notes for revision

PUBLISHED

PLUMBING
ISBN 0 340 71911 7

FORTHCOMING

BRICKWORK
ISBN 0 340 71913 3

CARPENTRY AND JOINERY
ISBN 0 340 71914 1

PAINTING AND DECORATING
ISBN 0 340 71912 5

PUBLISHED

ELECTRICAL INSTALLATION
THEORY AND PRACTICE
Maurice Lewis
ISBN 0 340 67665 5

ENGINEERING DRAWING FROM FIRST PRINCIPLES
USING AUTOCAD
Dennis Maguire
ISBN 0 340 69198 0

ELECTRONIC PRINCIPLES AND APPLICATIONS
John B Pratley
ISBN 0 340 69275 8

Revision and Self-Assessment Series

PLUMBING

KEVIN STEAD
on behalf of the Leeds College of Building

ELSEVIER
BUTTERWORTH
HEINEMANN

AMSTERDAM BOSTON HEIDELBERG LONDON NEW YORK OXFORD
PARIS SAN DIEGO SAN FRANCISCO SINGAPORE SYDNEY TOKYO

Elsevier Butterworth-Heinemann
Linacre House, Jordan Hill, Oxford OX2 8DP
200 Wheeler Road, Burlington, MA 01803

First published by Arnold 1988
Reprinted 1992, 2003 (twice)

British Library Cataloguing in Publication Data
A catalogue record for this book is available from the British Library

Library of Congress Cataloguing in Publication Data
A catalogue record for this book is available from the Library of Congress

ISBN 340 706309

Printed and bound in Malta by Gutenberg Press

CONTENTS

1

BACKGROUND TECHNOLOGY 1

To tackle this chapter you will need to know:

- how to calculate the area of common shapes found in plumbing systems;
- how to calculate the volumes of three-dimensional objects;
- how to calculate the capacities of these objects.

Area of a circle = πr^2
Area of an open-ended cylinder = $2\pi rh$
Area of a closed-ended cylinder = $2\pi r^2 + 2\pi rh$
Circumference of a circle = πd
Area of a square = length × length
Area of a rectangle = length × width
Area of a cube = 6(length × length)
Area of a cuboid = 2(length × width) + 2(width × depth) + 2(length × depth)
Volume of a cube = length × length × length
Volume of a cuboid = length × width × depth
Volume of a cylinder = $\pi r^2 h$
Capacity of a cuboid = length × width × depth × 1000
Capacity of a cylinder = $\pi r^2 h \times 1000$
Area of a triangle = 0.5 base × height

GLOSSARY OF TERMS

Area – a two-dimensional measurement that usually requires the multiplication of two numbers, e.g. length times width. Area is measured in the SI system as metres squared (m^2).
Capacity – the measurement of fluid contained within a shape. In plumbing the fluid is normally water and there are 1000 litres of water in 1 m^3 of volume. Therefore to calculate the capacity of a shape we simply multiply the volume by 1000.
Circle – a two-dimensional shape and therefore measured in square metres (m^2). The shape is said to be round.
Circumference – the line of the perimeter of a circle, a linear measurement (m).
Cylinder – a three-dimensional shape which is circular in cross-section with an unspecified length. A cylinder has a surface area measured in metres squared (m^2) and a volume which is measured in metres cubed (m^3).
Depth – a linear measurement that refers to the vertical measurement of a three-dimensional shape, such as a cube or cuboid.
Diameter – a straight line measured from a point on the circumference passing through the centre of the circle to the opposite part of the circumference (i.e. twice the radius).
Length – a linear measurement, i.e. a straight line. This is usually the longest dimension of a shape. Length is measured in metres (m) in the SI system.
Pi (π) – a constant-value number which is the relationship of the diameter of a circle to the circumference; of the radius squared to the area of a circle, etc.. The value of π to three decimal places is 3.142.
Radius – a straight line measured from the centre of a circle to the circumference. It is measured in metres (m).
Square – a four-sided two-dimensional shape whose length and width are exactly the same dimension. The unit of measurement is square metres (m^2).
Triangle – a three-sided shape which can be used in many types of calculations, e.g. the setting up of a right angle – the 5:4:3 rule.
Volume – a three-dimensional measurement of space within a three-dimensional shape, e.g. a cube or cylinder. Volume is expressed as m^3.
Width – a linear measurement which often refers to the shorter dimension of a two-dimensional shape. Width is measured in metres (m).

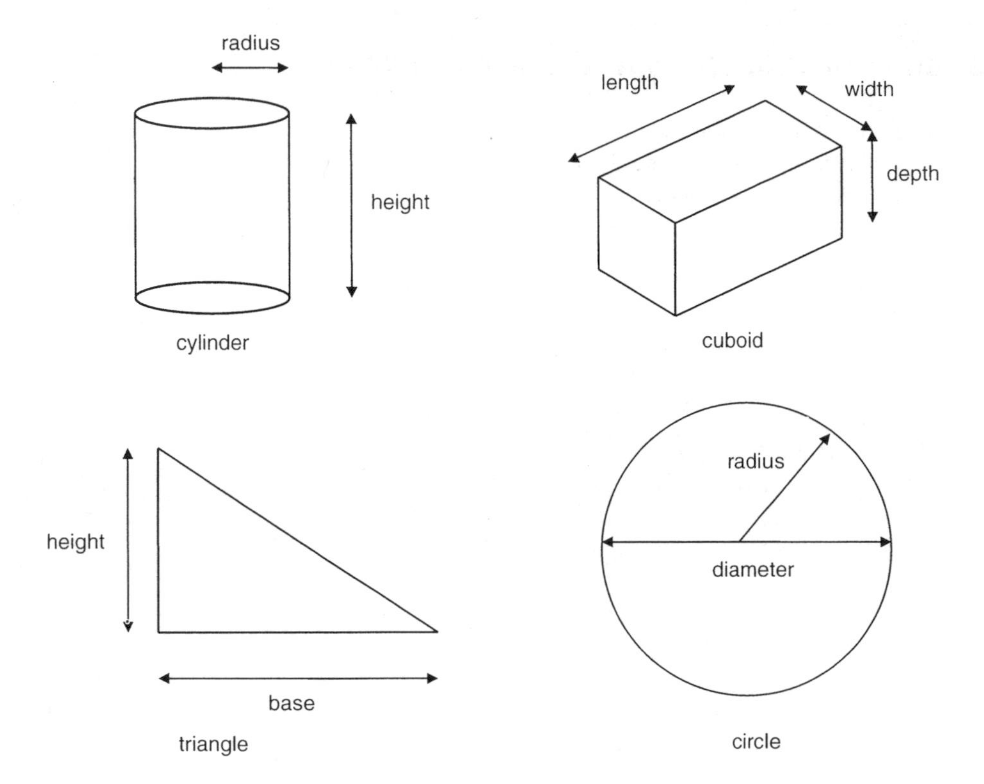

Fig. 1.1

Examples of calculations

Set out below are worked examples of the calculations in this chapter. If you set out your calculations to the same standard it should help to follow the process through stage by stage.

1. Calculate the area of a circle if the radius is 1 metre.

$$\begin{aligned}\text{Area} &= \pi r^2 \\ &= 3.142 \times 1 \times 1 \\ &= \underline{3.142\ \text{m}^2}\end{aligned}$$

2. Calculate the surface area of sheet copper needed to make an open-ended cylinder if the length is to be 1.2 m and the radius is 0.25 m.

$$\begin{aligned}\text{Area} &= 2\pi rh \\ &= 2 \times 3.142 \times \text{radius} \times \text{height} \\ &= 2 \times 3.142 \times 0.25 \times 1.2 \\ &= \underline{1.885\ \text{m}^2}\end{aligned}$$

3. Calculate the surface area of a closed-ended cylinder if the radius is 0.3 m and the length is 1.5 m.

$$\begin{aligned}\text{Area} &= 2\pi r^2 + 2\pi rh \\ &= (2 \times 3.142 \times 0.3 \times 0.3) + (2 \times 3.142 \times 0.3 \times 1.5) \\ &= 0.566 + 2.828 \\ &= \underline{3.394\ \text{m}^2}\end{aligned}$$

4. Calculate the circumference of a circle with a diameter of 2 m.

$$\begin{aligned}\text{Circumference} &= \pi d \\ &= 3.142 \times \text{diameter} \\ &= 3.142 \times 2 \\ &= \underline{6.284\ \text{m}}\end{aligned}$$

5. If the sides of a square are 2 m in length, what will be the surface area of the square?

$$\begin{aligned}\text{Area of a square} &= \text{length} \times \text{length} \\ &= 2 \times 2 \\ &= \underline{4\ \text{m}^2}\end{aligned}$$

6. Calculate the area of a piece of glass if the length is 1.5 m and the width is 0.6 m.

$$\begin{aligned}\text{Area of a rectangle} &= \text{length} \times \text{width} \\ &= 1.5 \times 0.6 \\ &= \underline{0.9\ \text{m}^2}\end{aligned}$$

7. Calculate the area of a cube if all the sides of the cube measure 1 metre in length.

$$\begin{aligned}\text{Area of a cube} &= 6(\text{length} \times \text{length}) \\ &= 6 \times (1 \times 1) \\ &= 6 \times 1 \\ &= \underline{6\ \text{m}^2}\end{aligned}$$

8. Calculate the area of a storage cistern (cuboid) if its measurements are as follows: length 1 metre, width 0.6 metres and depth 0.4 metres.

$$\begin{aligned}\text{Area} &= 2(\text{length} \times \text{width}) + 2(\text{width} \times \text{depth}) + 2(\text{length} \times \text{depth}) \\ &= 2(1 \times 0.6) + 2(0.6 \times 0.4) + 2(1 \times 0.4) \\ &= 1.2 + 0.48 + 0.8 \\ &= \underline{2.48\ \text{m}^2}\end{aligned}$$

9. Calculate the volume of a cube that measures 4 m along all its sides.

$$\begin{aligned}\text{Volume} &= \text{length} \times \text{length} \times \text{length} \\ &= 4 \times 4 \times 4 \\ &= \underline{64\ \text{m}^3}\end{aligned}$$

10. Calculate the volume of the storage cistern in Question 8.

$$\begin{aligned}\text{Volume} &= \text{length} \times \text{width} \times \text{depth} \\ &= 1.0 \times 0.6 \times 0.4 \\ &= \underline{0.24\ \text{m}^3}\end{aligned}$$

11. Calculate the volume of a cylinder which measures 1.2 m high by 0.225 m radius.

Volume = $\pi r^2 h$
= 3.142 × radius × radius × height
= 3.142 × 0.225 × 0.225 × 1.2
= <u>0.191 m³</u>

12. Calculate the capacity of the cuboid in Question 10.

Capacity = length × width × depth × 1000
= 1.0 × 0.6 × 0.4 × 1000
= 0.24 × 1000
= <u>240 litres</u>

13. Calculate the capacity of the cylinder in Question 11.

Capacity = $\pi r^2 h$
= 3.142 × radius × radius × height × 1000
= 3.142 × 0.225 × 0.225 × 1.2 × 1000
= 0.191 × 1000
= <u>191 litres</u>

14. If a right-angled triangle has a base length of 2 m and a vertical height of 3 m, what is the area of the triangle?

Area of a triangle = 0.5 × base × height
= 0.5 × 2 × 3
= <u>3 m²</u>

Where a calculation has brackets included e.g. 2(2 + 4), the calculation in the bracket must be done first and then the sum is multiplied by the 2 so that the answer equals 2 × 6 = 12. Also your answer should include the units and not just the numerical value e.g. 6 m^2.

Now try and answer these questions

1. A flat roof measuring 4 m by 2 m is to be covered with sheet lead. Calculate the surface area of the flat.

   a 8 m^2
   b 0.8 m^2
   c 80 m^2
   d 8.5 m^2.

2. A hot water storage cylinder is 1.04 m high and has a diameter of 0.45 m. Therefore the volume of the cylinder is

   a 0.651 m^3
   b 1.65 m^3
   c 1.66 m^3
   d 0.165 m^3.

3. Calculate the surface area of a storage cistern which measures 600 mm × 500 mm × 450 mm deep.

   a 1.55 m^2
   b 1.59 m^2
   c 2.20 m^2
   d 2.56 m^2.

4. A circular cold water storage cistern has a diameter of 0.8 m and the height to the waterline is 0.5 m. Calculate the capacity of the cistern in litres.

   a 251 litres
   b 185 litres
   c 261 litres
   d 158 litres.

5. If a cylinder has a diameter of 500 mm, what will be the minimum length of strap required to go round the cylinder?

   a 1.67 m
   b 1.57 m
   c 1500 mm
   d 1600 mm.

6. Calculate the length of 'gas identification tape' required to wrap around a 110 mm diameter gas main.

   a 0.436 m
   b 450 mm
   c 350 mm
   d 0.346 m.

7. If the circumference of a circle is 3m, what will be its diameter?

   a 0.955 m
   b 1.00 m
   c 970 mm
   d 1000 mm.

8. Calculate the circumference of a circle if its diameter is 1.2 m.

   a 4.77 m
   b 3.77 m
   c 2.77 m
   d 5.77 m.

9. The diameter of a flat-ended cylinder is 0.5 m. What will be its cross-sectional area?

   a 0.259 m^2
   b 1.59 m^2
   c 2.59 m^2
   d 0.159 m^2.

10. Calculate the surface area of the washer in the diagram in Fig. 1.2.

    a 0.412 m^2
    b 4.12 m^2
    c 0.0412 m^2
    d 41.20 m^2.

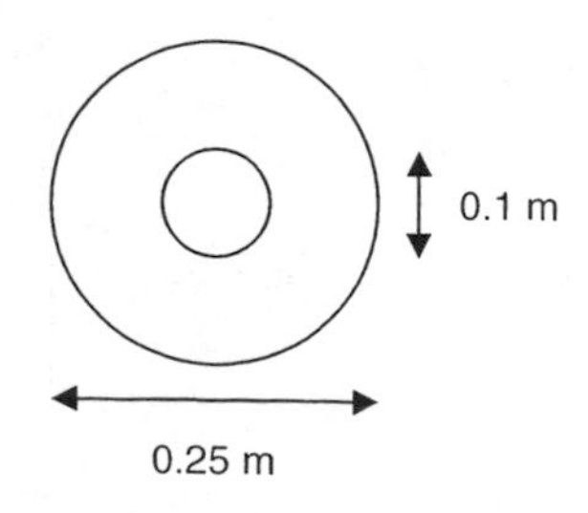

Fig. 1.2

11. Calculate the cross-sectional area of a 150 mm drainpipe.

a 0.0177 m^2
b 0.177 m^2
c 1.77 m^2
d 17.7 m^2.

12. If a door has a glass panel measuring 650 mm × 650 mm, and the glass costs £2.00 per square metre, calculate the area and the cost of the piece of glass.

a 0.450 m^2; £0.95
b 4.22 m^2; £8.50
c 0.5 m^2; £1.00
d 0.422 m^2; £0.85.

13. A window frame has 20 small pieces of glass, each measuring 200 mm × 200 mm. Calculate the total area of the glass in the frame.

a 0.8 m^2
b 0.9 m^2
c 8 m^2
d 9 m^2.

14. A bathroom is to be half tiled, the tiles chosen measure 150 mm × 150 mm. If the area to be tiled is 5 m^2, how many tiles will be required to tile the bathroom?

a 224 tiles
b 22.5 tiles
c 222 tiles
d 22.2 tiles.

15. Calculate the area of the triangle in Fig. 1.3.

a 1.0 m^2
b 0.5 m^2
c 1.5 m^2
d 0.75 m^2.

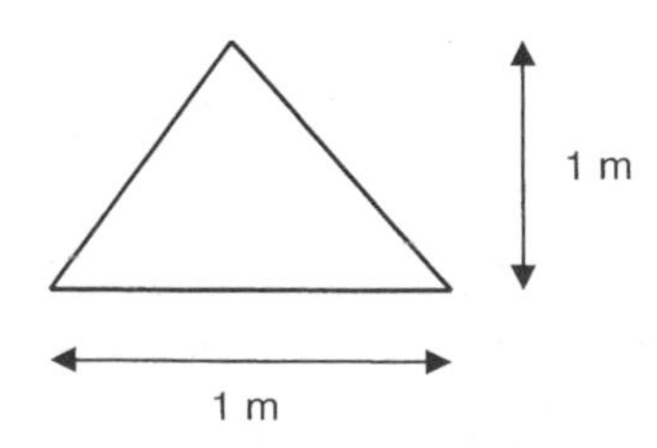

Fig. 1.3

16. Calculate the area of the triangle in Fig. 1.4.

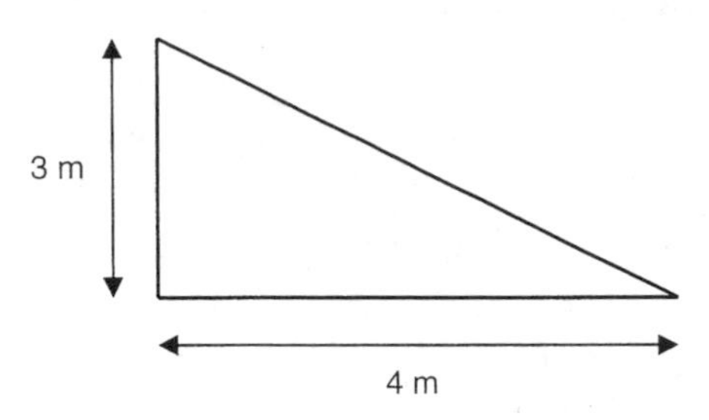

Fig. 1.4

a 6 m^2
b 16 m^2
c 3 m^2
d 12 m^2.

17. Calculate the area of the triangle in Fig. 1.5.

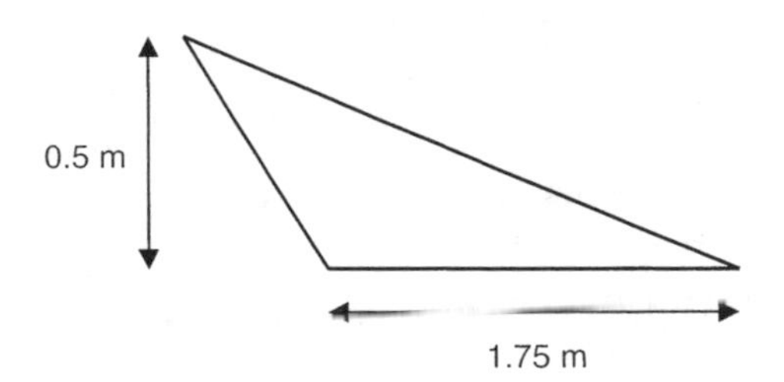

Fig. 1.5

a 0.438 m^2
b 0.45 m^2
c 4.38 m^2
d 0.0438 m^2.

18. A 6900 litre oil storage tank measures 3 m × 1.85 m × 1.25 m. Calculate the area of the steel sheet required to fabricate the tank.

a 2.322 m^2
b 25.50 m^2
c 23.225 m^2
d 30.00 m^2.

19. A block of cast iron measures 200 mm × 300 mm × 150 mm. What is its total surface area?

a 2.5 m^2
b 0.25 m^2

c $2.7\ m^2$
d $0.27\ m^2$.

20. A hot water storage tank is to be lagged with cork slabs on all six sides. If the tank measures 1.2 m high × 450 mm × 450 mm, how many square metres of cork need to be ordered?

a $28.65\ m^2$
b $2.865\ m^2$
c $2.565\ m^2$
d $25.65\ m^2$.

21. 500 litres of stored water are necessary for a building; the base provided measures 1 m × 0.8 m. What is the minimum height to the waterline of the cistern?

a 0.825 m
b 0.625 m
c 1.01 m
d 1.25 m.

22. A cube which has lengths of 1 m is to be totally covered in sheet lead. What is the minimum surface area of lead required?

a $1\ m^3$
b $1\ m^2$
c $6\ m^3$
d $6\ m^2$.

23. Calculate the surface area of a 150 mm cube in m^2.

a $0.135\ m^2$
b $0.250\ m^2$
c $1.35\ m^2$
d $2.5\ m^2$.

24. Calculate the amount of sheet steel required to make an oil tank measuring 2 m × 2 m × 2 m high.

a $2.4\ m^2$
b $25\ m^2$
c $24\ m^2$
d $30\ m^2$.

25. Calculate the volume of a cuboid if its physical dimensions are 4 m long × 3 m wide × 1 m high.

a $10\ m^3$
b $12\ m^3$
c $1.2\ m^3$
d $1.0\ m^2$.

26. A storage cistern measures 1.2 m in length × 0.6 m wide × 0.56 m high. Calculate the volume of the cistern.

a $0.403\ m^3$
b $0.413\ m^3$
c $0.603\ m^3$
d $0.613\ m^3$.

27. If the volume of a tank is $1.5\ m^3$ and the base measures 1 m × 0.5 m, what is the height of the tank?

a 3.5 m
b 2 m
c 2.5 m
d 3 m.

28. How many litres of water can be stored in a tank measuring 3 m × 2 m × 1 m high?

a 5000 litres
b 6000 litres
c 1600 litres
d 1200 litres.

29. A cold water storage cistern has the following dimensions: 0.8 m × 0.6 m × 0.45 m up to the the waterline. How many litres of water can be stored within it?

a 316 litres
b 112 litres
c 216 litres
d 250 litres.

30. Calculate the surface area of an open-ended cylinder measuring 2 m long with a diameter of 200 mm.

a $12.56\ m^2$
b $1.256\ m^2$
c $2.561\ m^2$
d $25.61\ m^2$.

31. A section of circular ventilation duct has a diameter of 350 mm and measures 10 m in length. Calculate the area of sheet steel required to fabricate the duct.

a 1.256 m^2
b 12.56 m^2
c 10.56 m^2
d 1.056 m^2.

32. The sleeve part of a lead slate is 200 mm long and has a diameter of 110 mm. Calculate the area of sheet lead required to make it.

a 0.069 m^2
b 0.69 m^2
c 0.12 m^2
d 0.012 m^2.

33. A hot water storage cylinder has a radius of 225 mm and a height of 1040 mm. What will be the volume of the cylinder?

a 1.65 m^3
b 0.25 m^3
c 0.165 m^3
d 2.5 m^3.

34. A hot water storage calorifier is 2.5 m high and 1.2 m diameter. Calculate the volume of the calorifier.

a 0.283 m^3
b 3.5 m^3
c 0.35 m^3
d 2.83 m^3.

35. A circular cold water storage vessel has a diameter of 600 mm and contains water 450 mm deep up to the waterline. What will be the volume of water stored?

a 1.27 m^3
b 0.127 m^3
c 2.17 m^3
d 0.217 m^3.

36. A cylinder has a volume of 0.5 m^3 and a radius of 0.3 m. How high is the cylinder?

a 0.577 m
b 2.77 m
c 1.77 m
d 1.07 m.

37. Calculate the surface area of a closed cylinder, which has a diameter of 1 m and a length of 2 m.

a 7.855 m^2
b 6.855 m^2
c 78.55 m^2
d 68.55 m^2.

38. A circular cold water storage cistern has a diameter of 600 mm and a height of 600 mm. Calculate the surface area of the cistern.

a 14.14 m^2
b 2.414 m^2
c 24.14 m^2
d 1.414 m^2.

39. Calculate the volume of the cylinder referred to in Question 37.

a 0.257 m^3
b 2.571 m^3
c 1.571 m^3
d 0.157 m^3.

40. Calculate the capacity of the cylinder referred to in Question 37.

a 1571 litres
b 157 litres
c 2571 litres
d 257 litres.

41. Calculate the volume of the cold water storage cistern referred to in Question 38.

a 1.697 m^3
b 0.1697 m^3
c 0.2697 m^3
d 2.697 m^3.

42. Calculate the capacity of the cold water storage cistern referred to in Question 38.

a 2697 litres
b 269.7 litres
c 169.7 litres
d 1697 litres.

Now check your answers from the grid.

Q 1; a	Q 10; c	Q 19; d	Q 28; b	Q 37; a
Q 2; d	Q 11; a	Q 20; c	Q 29; c	Q 38; d
Q 3; b	Q 12; d	Q 21; b	Q 30; b	Q 39; c
Q 4; a	Q 13; a	Q 22; d	Q 31; a	Q 40; a
Q 5; b	Q 14; c	Q 23; a	Q 32; a	Q 41; b
Q 6; d	Q 15; b	Q 24; c	Q 33; c	Q 42; d
Q 7; a	Q 16; a	Q 25; b	Q 34; d	
Q 8; b	Q 17; a	Q 26; a	Q 35; b	
Q 9; d	Q 18; c	Q 27; d	Q 36; c	

2

BACKGROUND TECHNOLOGY 2

To tackle this chapter you will need to know:

- the different scales used to measure temperature;
- the three methods of heat transfer;
- the different types of heat absorbed by substances;
- the different rate at which materials expand on heating or cooling;
- the relationship between heat input and temperature change;
- the different types of thermometers used to measure degrees or hotness.

GLOSSARY OF TERMS

Celsius – a temperature scale of 100 degrees: 0°C to 100°C, where 0°C is equivalent to 273 K and 100°C is equivalent to 373 K. This is the temperature scale used in the SI system. 0°C is the freezing point (FP) of water and 100°C is the boiling point (BP) of water at atmospheric pressure.

Coefficient of linear expansion – the fraction of a body's original length by which a substance expands per degree rise in temperature. A coefficient is a constant, which is to say it never varies. Coefficients are derived by experimentation and they are of particular value when assessing the movement of pipework which is subjected to temperature rise, as with heating system pipework, etc.

Conduction – the transfer of heat through a solid material such as metals or masonry.

Convection – the transfer of heat within a fluid, i.e. a gas or a liquid such as air and water. Heat is transferred within the mass of the fluid by part of it becoming less dense due to the heating effect and then being displaced to the top of the mass of fluid.

Expansion of water – the increase in volume of water with a change in temperature. Water is a substance, which like most others, will expand on being heated. But it will also expand when it is cooled below 4°C (maximum density of water). Generally the plumber is more concerned with the expansion of water due to heating, as in the case of either unvented hot water storage or heating systems and the potential for causing damage. The expansion of water on being heated can be found using the equation

$$E = V\left[\frac{\rho_1 - \rho_2}{\rho_2}\right]$$

where:

E = expansion of water (m^3)
V = volume of water before being heated (m^3)
ρ_1 = density of water before being heated (kg/m^3)
ρ_2 = density of water after being heated (kg/m^3)

Table 2.1 Coefficient of linear expansion

Common materials	Coefficient of linear expansion
Aluminium	0.000 023
Brass	0.000 019
Copper	0.000 016
Iron	0.000 01
Lead	0.000 029
Steel – low carbon	0.000 011
Steel – invar	0.000 001
Tin	0.000 021
Zinc	0.000 026
Polyethylene – low-density	0.000 28
Polyethylene – high-density	0.000 11
Polyvinyl – normal-impact	0.000 05
Polyvinyl – high-impact	0.000 08

Table 2.2 Density of water at various temperatures

Water temperature (0°C)	Water density (kg/m^3)	Water temperature (0°C)	Water density (kg/m^3)
0	999.8	60	983.0
4	1000	63	982.0
10	999.7	65	980.7
15.6	999.0	66	980.1
20	998.0	70	977.5
25	996.0	71	977.0
30	995.0	74	975.0
38	993.0	80	972.0
40	992.0	82	970.4
46	989.0	91	965.0
50	987.5	100	958.0

Fahrenheit – the scale used in pre-metrication days and has a freezing point for water of 32°F and a boiling point of 212°F at atmospheric pressure.

Heat – a form of energy. Energy cannot be created or destroyed, but it can be converted from one form to another e.g. from electrical energy into heat as in the boiling of water in a kettle. Heat is measured in joules (J).

Heat quantity – the product of three factors: the mass of a substance, the specific heat capacity of a substance and the rise or fall in the substance's temperature. It can be used to determine the energy requirements for heating a substance or conversely the amount of heat retrievable from a substance by cooling it.

Heat transfer – the movement of heat from one body to another, i.e. from a hot body to a cold body. Heat transfer is driven by a difference in temperature, thus the greater the difference in temperature the more intense the rate of transfer. Heat transfer can be in three different forms: conduction, convection or radiation.

Kelvin – often referred to as 'absolute temperature', the scale is from 0 K to infinity. Kelvin is used for scientific calculations.

Latent heat – the heat absorbed by a substance which will bring about a 'change in state' without a change in temperature; for instance, ice to water.

Radiation – the transference of heat through space or a clear gas by means of electromagnetic waves. The classic example is the heat received on the earth's surface from the sun.

Temperature – a measure of hotness or coldness of a substance. Temperature is measured in degrees by an instrument called a thermometer, typical scales on thermometers being Kelvin, Celsius or Fahrenheit. Temperature is not a measure of heat but something which is brought about by the absorption or rejection of heat in a substance.

Thermal conductors – Substances that conduct heat. Solid substances are either good or poor conductors of heat. A good conductor for example is a metal such as copper and an example of a poor conductor might be a piece of wood. Metal substances tend to be good conductors.

Thermometer – an instrument for measuring degrees of hotness, and will sense both a rise or fall in temperature. It is not capable of measuring heat quantity directly. There are many types of thermometer available; the most common being the mercury or alcohol in glass tube type. It is also possible to obtain electronic instruments that will measure either a single temperature or two temperatures simultaneously – these are useful when commissioning heating systems for example.

Total heat – the sum of both the latent and the sensible heat of a substance. Total heat is sometimes referred to as enthalpy, particularly in steam tables.

Sensible heat – the heat absorbed by a substance which will bring about a 'change in temperature'. For example, when water is heated up in a kettle from say 10°C to 100°C to make tea.

Specific heat capacity (SHC) – a substance's capacity for absorbing heat. It is defined as 'the amount of heat required to raise 1 kg of a substance through 1°C', e.g. water has an SHC of 4.2 kJ/kg/°C. All substances have a specific heat capacity; for some common materials refer to Table 2.3. The heat absorbed can also be retrieved for future use. Water is used to transport heat through pipes from the boiler to the radiators in a heating system – it has a high SHC.

Table 2.3 Specific heat capacity

Common materials	Specific heat capacities (kJ/kg/°C)
Mercury	0.14
Iron	4.6
Copper	0.4
Lead	0.13
Brick	8.4
Concrete	9.2
Glass	6.7
Ice	2.1
Water	4.2 (4.186)
Air	1.0 (1.012)
Methylated spirit	2.5

Watt – the unit of power in the SI system and may be expressed as the relationship between energy and time. Thus 1 watt is equal to 1 joule per second. The amount of energy represented by 1 Watt is very small, therefore power is often measured in units of 1000 watts, or 1 kilowatt (kW).

Examples of equations

HEAT QUANTITY

$$Q = M \times \text{SHC} \times (t_2 - t_1)$$

where:

Q	=	heat quantity	(kJ)
M	=	mass of a substance	(kg)
SHC	=	specific heat capacity of a substance	(kJ/kg)
t_1	=	the temperature before heating	(°C)
t_2	=	the temperature after heating	(°C)

EXPANSION OF MATERIALS

$$E = L \times C \times (t_2 - t_1)$$

where:

E	=	rate of expansion	(m)
L	=	original length of the material	(m)
C	=	coefficient of linear expansion	(m/°C)
t_1	=	the temperature before heating	(°C)
t_2	=	the temperature after heating	(°C)

POWER

$$P = \frac{\text{heat quantity}}{\text{time}} = \frac{Q}{t}$$

where:

P	=	watts or kilowatts	(W) or (kW)
Q	=	heat quantity	(J) or (kJ)
t	=	time	(s)

TEMPERATURE

Conversion from Fahrenheit to Celsius

$$°\text{C} = \frac{5}{9}\,(°\text{F} - 32)$$

Conversion from Celsius to Fahrenheit

$$°\text{F} = \frac{9 \times °\text{C}}{5} + 32$$

To convert from Celsius to Kelvin

$$°\text{C} + 273$$

To convert from Kelvin to Celsius

$$\text{K} - 273$$

Worked examples of the equations

Example 2.1 Calculate the expansion volume of a domestic heating system with a cold volume of 60 litres of water, which has a temperature of 10°C. When the system is running normally its mean temperature is 80°C.

$$E = V\left[\frac{\rho_1 - \rho_2}{\rho_2}\right]$$

$$\text{Expansion} = 0.06 \times \frac{[999.7 - 972]}{972}$$

$$= 0.06 \times 0.028$$

$$= \mathbf{1.68\ litres}$$

Example 2.2 A hot water storage cylinder has a volume of 119 litres of water at 10°C. It is required to raise its temperature to 65°C. Calculate the heat required in kJ to bring about this change in temperature. Take it that 1 litre of water is equivalent to 1 kg.

$$\begin{aligned} Q &= M \times \text{SHC} \times (t_2 - t_1) \\ &= 119 \times 4.2 \times (65 - 10) \\ &= 119 \times 4.2 \times 55 \\ &= \underline{27\,489\text{ kJ}} \end{aligned}$$

Example 2.3 A length of copper rod has a mass of 10 kg and it is at a temperature of 5°C. Calculate the amount of heat required to raise its temperature to 90°C.

$$\begin{aligned} Q &= M \times \text{SHC} \times (t_2 - t_1) \\ &= 10 \times 0.4 \times (90 - 5) \\ &= 10 \times 0.4 \times 85 \\ &= \underline{340\text{ kJ}} \end{aligned}$$

Example 2.4 A storage cylinder has a capacity of 150 litres. If the cold temperature is 10°C and the hot teperature is 60°C and the water is to be heated up in two hours calculate the power rating of an immersion heater capable of heating the water.

$$\begin{aligned} Q &= M \times \text{SHC} \times (t_2 - t_1) \\ &= 150 \times 4.2 \times (60 - 10) \\ &= 150 \times 4.2 \times 50 \\ &= \underline{31\,500\text{ kJ}} \end{aligned}$$

$$\begin{aligned} \text{Power rating} &= \frac{\text{Heat quantity}}{\text{time}} \\ &= \frac{31\,500}{7200} \\ &= \underline{4.375\text{ kW}}, \quad \underline{\text{say } 4.5\text{ kW heater}}. \end{aligned}$$

Example 2.5 A copper heating flow pipe is 20 m long. When cold the water inside it is 10°C. If the water is heated to 85°C when the system is operational calculate the amount of expansion that will take place in the pipe.

$$\begin{aligned} E &= L \times C \times (t_2 - t_1) \\ &= 20 \times 0.000\,016 \times (85 - 10) \\ &= 20 \times 0.000\,016 \times 75 \\ &= \underline{0.024\text{ m or }24\text{ mm}} \end{aligned}$$

Example 2.6 If a low-density polyethylene pipe was installed in the Example 2.5, compare the rate of expansion.

$$\begin{aligned} E &= L \times C \times (t_2 - t_1) \\ &= 20 \times 0.000\,28 \times (85 - 10) \\ &= 20 \times 0.000\,28 \times 75 \\ &= \underline{0.42\text{ m or }420\text{ mm}} \end{aligned}$$

Relationship of heat input to temperature change

Figure 2.1 demonstrates the relationship between the heat input applied to water and the rise in temperature. As can be seen there are changes in state taking place. This graph represents what occurs at atmospheric pressure. If the pressure exerted upon the water was altered at regular intervals of time a slightly different curve would occur. An increase in the pressure exerted, for example, would elevate the boiling point of water above 100°C. This is clearly demonstrated on the pressure/enthalpy graph used by steam engineers.

As heat is applied along the *X* axis from left to right the rise in temperature is plotted along the graph. The diagonal rising line (or curve) shows a rise in temperature; this is indicative of sensible heat being absorbed. There are three types of sensible heat: sensible heat of ice, water and steam.

Where the curve of the graph remains at the same temperature as more heat is applied, this is indicative of latent heat being absorbed by the water. The temperature remains constant until a complete change in state has occurred. There are two types of latent heat: latent heat of fusion (ice) and latent heat of evaporation (or condensation).

It is possible for water to exist in two different states simultaneously, i.e. water and ice, or water and steam. This indicates that not all the latent heat has been absorbed or conversely given up. As can be seen water as ice contains a small amount of heat down to as low as −272°C (1 K).

This graph can be linked to the working principles of several plumbing systems, particularly domestic hot water supply and heating.

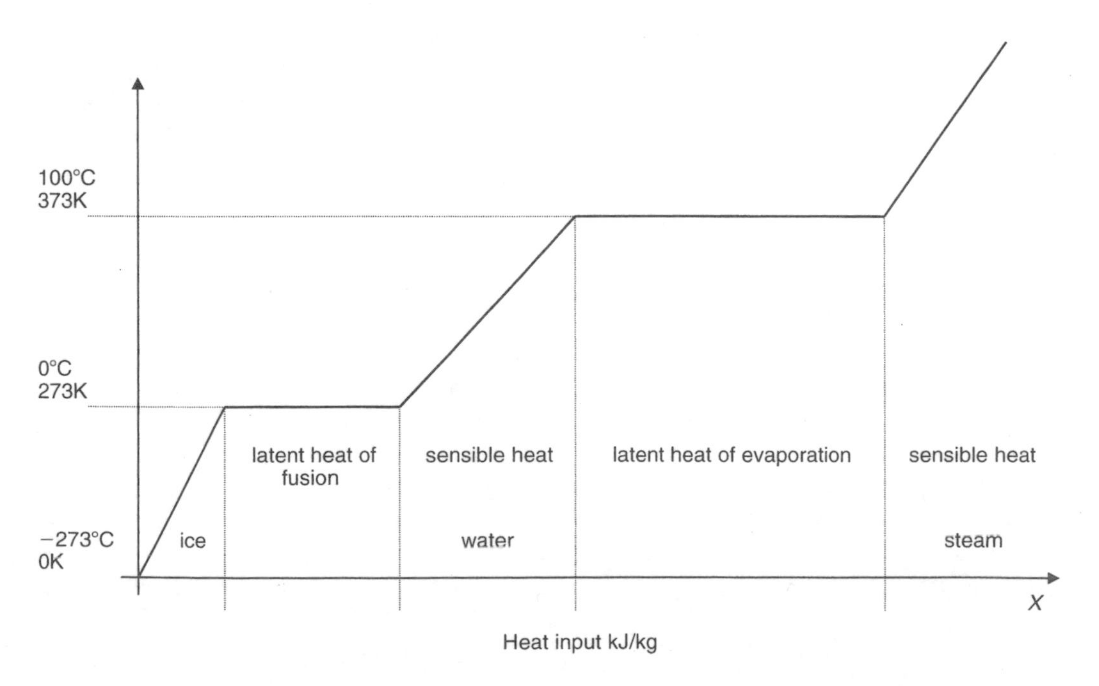

Fig. 2.1

Now try and answer these questions

1. The method of heat transfer which requires a solid material to act as a medium is referred to as
   - a conduction
   - b convection
   - c radiation
   - d propulsion.

2. If the temperature of a volume of water is at 180°F, what is the temperature in degrees Celsius?
   - a 100°C
   - b 72.2°C
   - c 32°C
   - d 82.2°C.

3. In the SI system of measurement the unit of power is measured in
   - a ohms
   - b watts
   - c volts
   - d newtons.

4. The heat required to raise 50 litres of water through 60°C is
   - a 9500 kJ
   - b 15 600 kJ
   - c 12 600 kJ
   - d 20 000 kJ

5. A thermometer is an instrument for measuring
   - a heat quantity
   - b degrees of hotness
   - c specific heat capacity
   - d sensible heat.

6. Sensible heat is the heat energy which brings about a
   - a change in state
   - b change in density
   - c change in volume
   - d change in temperature.

7. A length of steel pipe is 50 m long and is heated through 60°C. The amount of expansion will be
   - a 33 mm
   - b 63 mm
   - c 1 mm
   - d 100 mm.

8. 'Absolute temperature' is measured in degrees
   - a Celsius
   - b Centigrade
   - c Kelvin
   - d Fahrenheit.

9. An unvented hot water cylinder has a volume of 0.165 m^3 at 10°C. If the immersion heater raises its temperature to 60°C, the expansion volume will be
   - a 2.803 litres
   - b 28.03 litres
   - c 12.00 litres
   - d 1.200 litres.

10. If the air temperature in a room is 21°C, the equivalent temperature in Fahrenheit is
    - a 80°F
    - b 50°F
    - c 69.8°F
    - d 79.8°F.

11. The quantity of heat contained in a substance is the product of
    - a two factors
    - b three factors
    - c four factors
    - d five factors.

12. Latent heat is the heat energy which brings about a
    - a change in state
    - b change in colour

c change in content
d change in temperature.

13. A coefficient is a __________, which is to say it never varies.

a variable
b theoretical number
c technical number
d constant.

14. Heat energy in water within a radiator will transfer through the wall of the radiator to the surrounding air by

a radiation
b convection
c conduction
d penetration.

15. A central heating system has a volume of 100 litres. If the temperature rise is from 10°C to 82°C, the expansion volume will be

a 3.02 litres
b 6.04 litres
c 4.08 litres
d 30.2 litres.

16. If pressure is exerted on the surface of water it will cause the __________ to rise.

a evaporation rate
b boiling point
c water level
d condensation.

17. Heat transfer will only take place when there is a

a difference in temperature
b difference in density
c lack of insulation
d clear passage.

18. The heat required to warm up a cylinder is 27 500 kJ. What size of heater is required to warm it up in two hours?

a 6.51 kW
b 1.5 kW
c 2.0 kW
d 3.82 kW.

19. If 10 kg of water contains 21 kJ of latent heat and 840 kJ of sensible heat, the total heat content of the water will be

a 40 kJ
b 17 640 kJ
c 819 kJ
d 861 kJ.

20. Specific heat capacity is a substance's capacity to

a convert heat
b transfer heat
c absorb heat
d produce heat.

21. The coefficient of linear expansion for low-carbon steel is

a 0.000 001 m/°C
b 0.000 011 m/°C
c 0.000 01 m/°C
d 0.000 11 m/°C.

22. The density of water at 100°C is

a 958.0 kg/m^3
b 980.1 kg/m^3
c 993.0 kg/m^3
d 999.8 kg/m^3.

23. Radiation is the transfer of heat through

a space
b water
c concrete
d wood.

24. The specific heat capacity of copper is

a 4.2 kJ/kg/°C
b 2.1 kJ/kg/°C
c 0.4 kJ/kg/°C
d 1.0 kJ/kg/°C.

25. A hot water storage cylinder contains 250 litres of water and is heated through 60°C. What size heater is required to heat the water in 1 hr?

a 1.75 kW
b 17.5 kW
c 3.5 kW
d 35 kW.

26. The maximum density of water occurs when its temperature is

a 0°C
b −4°C
c 10°C
d 4°C.

27. Convection is a type of heat transfer which requires a __________ to act as a medium.

a fluid
b solid
c space
d conduit.

28. Heat transfer by radiation is transmitted by

a a fluid
b a solid
c a liquid
d electromagnetic waves.

29. If a substance is at a temperature of 100°C, its absolute temperature will be

a − 373 K
b 373 K
c 273 K
d 0 K.

30. Absolute zero temperature (0 K) can also be expressed as __________.

a −100°C
b 273°C
c −273°C
d 100°C.

31. A volume of water has a temperature of 110° Fahrenheit. Measured in Celsius its temperature would be

a 43.3°C
b 53.3°C
c 120°C
d 212°C.

32. The most common type of thermometer is the

a absolute type
b electronic type
c glass tube type
d pyrometer type.

33. The heat absorbed by water which changes it into steam is said to be the latent heat of

a steam
b water
c fusion
d evaporation.

34. An electric storage heater has a 40 kg mass of concrete as a heat store. If its temperature is raised by 50°C the stored heat will be

a 18 400 kJ
b 14 800 kJ
c 6500 kJ
d 50 000 kJ.

35. When heat is absorbed by a substance it will bring about a

a rise in temperature
b fall in temperature
c change in colour
d change in use.

36. A PVC gutter installed around a building has a total length of 50 m. If it experiences a temperature change of 20°C the rate of expansion will be

a 5 mm
b 15 mm
c 55 mm
d 50 mm.

37. A square metre of brick wall has a mass of 242 kg. If the temperature of the wall is raised by 11°C the heat absorbed will be

a 11 360 kJ
b 22 360 kJ
c 20 360 kJ
d 15 360 kJ.

38. If a block of ice has a temperature of 25°F, the temperature in Celsius would be

a 3.89°C
b − 3.89°C
c − 1°C
d 1°C.

39. A central heating system has a water volume of 200 litres. The cold temperature is 10°C and its operating temperature is 82°C. The expansion volume on heating will be

a 16.03 litres
b 0.06 litres
c 6.03 litres
d 60 litres.

40. A cast iron boiler has a mass of 60 kg and contains 40 litres of water at 10°C. Calculate the heat required to raise the temperature of both to 85°C.

41. Calculate the size of immersion heater required to heat the contents of a 1040 mm × 450 mm diameter cylinder from 10°C to 60°C in a period of 1½ hours.

42. A hot water storage cylinder installed in a hotel has a capacity of 2000 litres. What size of heater (boiler) is required to heat the water through 55°C within a 2 hour period?

43. A section of lead flashing has a continuous length of 6 m. If the temperature change during a 24 hour period is 18°C, calculate the rate of expansion experienced by the lead.

44. A low carbon steel heating main is 50 m long and is subjected to a 75°C temperature rise. What will be the rate of expansion?

45. If 100 litres of water at 10°C has 16 800 kJ of heat applied to it, what will be the final temperature of the water?

Now check your answers from the grid.

Q 1; a	Q 10; c	Q 19; d	Q 28; d	Q 37; b
Q 2; d	Q 11; b	Q 20; c	Q 29; b	Q 38; b
Q 3; b	Q 12; a	Q 21; b	Q 30; c	Q 39; c
Q 4; c	Q 13; d	Q 22; a	Q 31; a	
Q 5; b	Q 14; c	Q 23; a	Q 32; c	
Q 6; d	Q 15; a	Q 24; c	Q 33; d	
Q 7; a	Q 16; b	Q 25; b	Q 34; a	
Q 8; c	Q 17; a	Q 26; d	Q 35; a	
Q 9; a	Q 18; d	Q 27; a	Q 36; d	

MODEL ANSWERS

Question 40

i $Q_{iron} = M \times \text{SHC} \times (t_2 - t_1)$
$60 \times 4.6 \times (85 - 10)$
$60 \times 4.6 \times 75$
$\underline{20\,700 \text{ kJ}}$

ii $Q_{water} = M \times \text{SHC} \times (t_2 - t_1)$
$40 \times 4.2 \times (85 - 10)$
$40 \times 4.2 \times 75$
$\underline{12\,600 \text{ kJ}}$

iii Total heat $= 20\,700 + 12\,600 = \underline{33\,300 \text{ kJ}}$

Question 41

i Capacity of cylinder $= \pi r^2 \times h \times 1000$
$= 3.142 \times 0.225^2 \times 1.04 \times 1000$
$= \underline{165 \text{ litres}}$

ii Heat required $= M \times \text{SHC} \times (t_2 - t_1)$
$= 165 \times 4.2 \times (60 - 10)$
$= 165 \times 4.2 \times 50$
$= \underline{34\,650 \text{ kJ}}$

ii Power required $= \dfrac{\text{Heat quantity}}{\text{seconds}}$

$= \dfrac{34\,650}{5400}$

$=$ <u>6.42 kW</u>

Question 42

i Heat required $= M \times \text{SHC} \times (t_2 - t_1)$
$= 2000 \times 4.2 \times 55$
$=$ **<u>462 000 kJ</u>**

ii Power required $= \dfrac{\text{Heat quantity}}{\text{seconds}}$

$= \dfrac{462\,000}{(2 \times 3600)}$

$=$ <u>64 kW</u>

Question 43

$$E = L \times C \times (t_2 - t_1)$$
$$= 6 \times 0.000\,029 \times 18$$
$= 0.003\ 13$ **m or 3.13 mm**

Question 44

$$E = L \times C \times (t_2 - t_1)$$
$$= 50 \times 0.000\,011 \times 75$$
$=$ <u>0.0413 m or 41.3 mm</u>

Question 45

$$Q = M \times \text{SHC} \times (t_2 - t_1)$$

$$\frac{Q}{(M \times \text{SHC})} = t_2 - t_1$$

$$\frac{Q}{(M \times \text{SHC})} + t_1 = t_2$$

$$\frac{16\,800}{100 \times 4.2} + 10 = t_2$$

$40 + 10 = t_2$
<u>50°C</u> $= t_2$

3

BACKGROUND TECHNOLOGY 3

To tackle this chapter you will need to know:

- the concept of atmospheric pressure;
- the method of calculating force and the units used;
- the method of calculating pressure exerted by mass;
- the hydraulic jack principle.

Standard conditions

An atmosphere consisting mainly of oxygen (20.9%) and nitrogen (79.1%) – air – surrounds the earth. Air has a mass of approximately 1.02 kg/m^3 at sea level. Due to the earth's gravitational pull and the mass of the air a pressure is exerted – this we refer to as atmospheric pressure. Changes in air pressure can affect the weather, i.e. high pressure usually means good weather, low pressure is associated with stormy wet weather. Differences in air pressure can adversely affect a gas soundness test. Fortunately air pressure usually changes very gradually.

The intensity of air pressure diminishes the higher it is measured above sea level, which is where it is at its greatest – at sea level it measures 1.013 bar. In addition the temperature and the moisture content of the air will determine its actual mass. Hot, dry air will have a smaller mass than air which is both cold and heavily saturated. For reference purposes air is given standard conditions – standard reference conditions (src). These standard conditions are

15°C   1013 mbar   dry

1013 mbar is the average atmospheric pressure and 15°C is the average temperature.
A volume measured under standard conditions is indicated by st in brackets following the volume abbreviation, for example;

$$1\ m^3\ (st)$$

Force and pressure

FORCE

The unit of force in the SI system is the newton, named after Sir Isaac Newton. It is used for calculating pressure, stress, work and power. A force of 1 newton (N) is defined as that force which gives a body having a mass of 1 kilogramme an acceleration of 1 metre per second squared

$$\text{Force} = \text{Mass} \times \text{Acceleration}$$

where:

Mass	=	the amount of matter measured	(kg)
Acceleration	=	is the acceleration of the mass	(m/s^2)

Note: Acceleration due to the earth's gravity is 9.81 m/s^2.

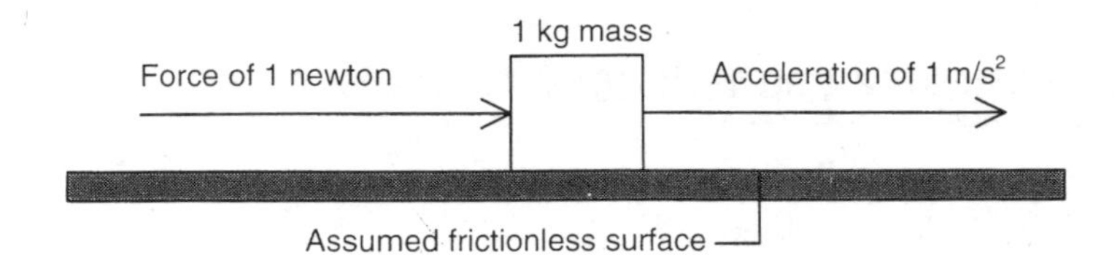

Fig. 3.1 Newton of force

For example if a mass of 1 kg was placed on a table it would exert a force of

Force = mass × acceleration
=1 × 9.81
=9.81 newtons

or alternatively if a mass of 20 kg was moved along at 3 m/s on a frictionless surface the force needed to be applied would be

Force = mass × acceleration
= 20 × 3
= 60 newtons

Now try a few examples for yourself.

1. A block of metal has a mass of 100 kg and it sits on a surface. What force is exerted by the block?

(981 newtons)

2. A boiler with a mass of 165 kg is sited on a concrete plinth. What force is applied by the boiler?

(1618.65 newtons)

3. What force would have to be applied to the boiler above to move it at 0.1 m/s?

(16.5 newtons)

PRESSURE

$$\text{Pressure} = \frac{\text{force}}{\text{area}}$$

where:

force = newtons (N)
area = unit area (m^2)

Therefore pressure is measured in newtons per metre squared (N/m^2). Remember also that 1 N/m^2 is equivalent to 1 pascal (Pa). The pascal is a unit of pressure used in the SI system, and is used in calculating pressure loss in pipes and ducts.

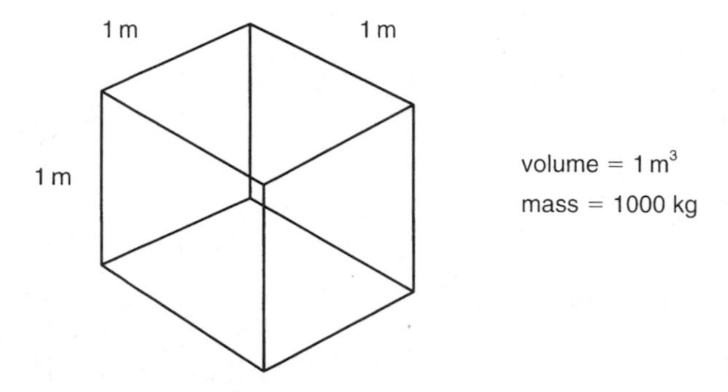

Fig. 3.2 Pressure on the base of a cubic metre of water

Now, as an example, let us consider an imaginary volume of water of 1 m^3 in size. The recognized density of water for simple calculations is 1000 kg/m^3. The force exerted by the water is

$$\begin{aligned}\text{Force} &= \text{mass} \times \text{acceleration}\\ &= 1000 \times 9.81\\ &= \underline{9810 \text{ newton} \quad \text{or} \quad 9.81 \text{ kilonewtons}}\end{aligned}$$

(the kilonewton is the preferred unit, as 1 N is a very small amount of force).
Since the force is acting over 1 square metre, the pressure on the base will be

$$\begin{aligned}\text{Pressure} &= \frac{\text{force}}{\text{area}}\\ &= \frac{9801}{1}\\ &= \underline{9810 \text{ N/m}^2}\end{aligned}$$

or preferably

$$\underline{9.81 \text{ kN/m}^2}$$

Since the pressure is proportional to the head of water, the pressure of water is calculated using

$$\text{Pressure} = 9.81 \text{ kN/m}^2 \times \text{Head of water (in m)}$$

In other words, since the column of water in the cubic metre was 1m high and gave us the pressure of 9.81 kN/m^2 it follows that the same value will be exerted by every metre of water height, so therefore the head of water is added to the calculation.

Example 3.1 A column of water is 10.33 m in height. Calculate the pressure on the base.

Pressure = 9.81 kN/m^2 × Head
= 9.81 × 10.33
= 101.33 kN/m^2

Note: The pressure of any liquid can be calculated using:

Pressure = Density × Acceleration × Head

If you think you understand this concept try a few examples yourself.

1. A column of water is 4.5 m high acting on a surface area of 1 m^2. What is the pressure exerted?

(44.15 kN/m^2)

2. A tank measures 2 m × 1.5 m × 2 m high. When full the water is 1.8 m deep. What pressure is acting on the base?

(17.66 kN/m^2)

How did you find those examples?

Now let us try a different liquid; consider an oil tank filled with oil, of density 900 kg/m^3, to 2 m. Calculate the pressure on the base.

Pressure = Density × Acceleration × Head
= 900 × 9.81 × 2
= 17 658 N/m^2 or 17.66 kN/m^2

Have you noticed, the pressure of 2 m of oil is the same as 1.8 m of water? Can you see how density of a substance affects pressure?

Carry on with the next problems.

3. Calculate the pressure of water acting upon a valve when the head of water above the valve is 15 m.

(147.15 kN/m^2)

4. The pressure acting on a boiler base is 210.92 kN/m^2. How high is the feed tank above the boiler?

(21.5 m)

5. A 25 mm gate valve in a pipe is holding back a head of water 8 m high. Calculate the pressure of water and also the force (N) applied to the disc of the valve.

i

($78.48\ kN/m^2$)

ii

(38.5 N)

You may require a little help with the last one, but try and sort it out for yourself first.

Absolute pressure

Atmospheric pressure, as we have seen, is 101.3 kN/m^2 at sea level. This value diminishes the higher you get above sea level.

Although not part of the SI system the 'bar' is a pressure value that is used frequently in building services systems.

Let us compare some different values:

The 'bar' is equal to 100 000 N/m^2 or Pascals (Pa)
which is 100 kN/m^2 or 100 kPa
Therefore atmospheric pressure is = 1.013 bar
100 N/m^2 = 1 mbar (used to measure low-pressure gas)
1 mbar will support a column of water approx 10 mm high.

When dealing with fluids such as refrigerants and steam, for instance, we must think in terms of absolute pressure.

Absolute pressure = gauge pressure + atmospheric pressure

The reason for this is that with thermal fluids, as mentioned above, pressure and temperature are fixed to one another, i.e. water boils at 100°C at atmospheric pressure and at no value of gauge pressure.

Gauge pressure is any pressure above atmospheric pressure. When a gauge is reading 0 bar it is in fact reading atmospheric pressure – 1.013 bar.

Therefore when a boiler gauge reads 2 bar it is actually reading 2 + 1.013 or 3.013 bar absolute. For simple calculations atmospheric pressure is taken as 101 kN/m^2 or 101 kPa.

Let us now consider a calculation.

Example 3.2 Calculate the absolute pressure acting on top of a boiler when the head of water above the top is 20 m.

Gauge pressure = 9.81 kPa × Head
= 9.81 × 20
= 196.2 kPa

Absolute pressure = 196.2 + 101
= 297.2 kPa abs

Now try a few calculations for yourself.

1. What is the absolute pressure exerted on a calorifier fed from a cistern 10 m above?

(199.1 kPa abs)

2. A reservoir is 100 m above a watermain. What is the absolute pressure on the pipe walls?

(1082 kPa abs)

3. Calculate the absolute pressure on a boiler if the head of water above it is 12.5 m?

(223.63 kPa abs)

Hydraulic press or jack

The hydraulic press is a machine which consists of two different-size pistons enabling it to lift weights by the application of a small force.

The weight that can be lifted depends on the ratio of the areas of the two pistons.

The force acting on each piston is equal to the pressure multiplied by the area.

$$\text{The ratio of piston area} = \frac{\text{large area}}{\text{small area}}$$

In Fig 3.3 the ratio is 0.1 divided by 0.01 which gives 1:10.

Since water is virtually incompressible, the force acting upon the larger piston will be ten times the force acting on the smaller piston.

$$\text{Therefore force applied} = 10 \text{ kN}$$

$$\text{force acting to lift} = 10 \text{ kN} \times 10 = 100 \text{ kN}$$

See Fig. 3.3.

The pressure applied to the water in the system is

$$\text{Pressure} = \frac{\text{force}}{\text{area}}$$

$$= \frac{10}{0.01}$$

$$= 1000 \text{ kN/m}^2$$

The pressure transmitted to the ram is

$$\text{Pressure} = \frac{\text{force}}{\text{area}}$$

$$1000 = \frac{\text{force}}{0.1}$$

Transposing the formula for force gives

$$\text{force} = \text{pressure} \times \text{area}$$
$$= 1000 \text{ x } 0.1$$
$$= \underline{100 \text{ kN}}$$

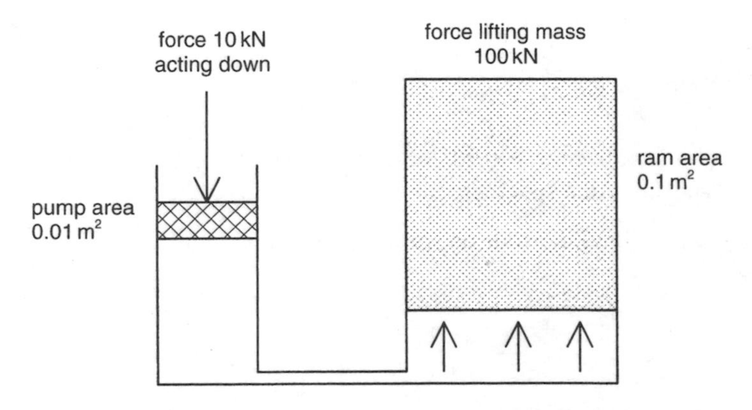

Fig. 3.3 Hydraulic principle

4

PIPEWORK

To tackle this chapter you will need to know:

- the materials used to manufacture pipes;
- the methods used to join pipes;
- the spacing of fixings;
- the type of fittings used on pipework;
- components used to control flow;
- the types of fixings used on pipes.

GLOSSARY OF TERMS

Bronze welding – a form of welding that requires a higher level of skill. The pipe ends require manipulation to producing a surface to weld on to. The welding alloy is manufactured from copper, zinc and nickel and therefore the process is not autogenous, but more of a soldering technique which requires a flux to assist the process. Oxyacetylene welding equipment is necessary to carry out this type of joint. Bronze welding is often used on large-diameter copper tubes such as sanitation pipework, etc.
Capillary joint – a method of soldering achieved by inserting a pipe end into a socket and melting solder into it. The gap between pipe end and socket is minimal so as to promote full penetration of molten solder throughout the joint. The soldering technique used is categorized as either hard or soft soldering. The joint is made by heating the pipe around the joint to the melting point of the solder being used. Solder is then applied to the socket. The solder melts and is taken into the joint by capillary attraction created by the two surfaces.
Copper pipe – copper pipe is manufactured to BS 2871: Part 1. It is produced in four common grades or tables within the standard. It is supplied in 3 m and 6 m straight lengths or coils in sizes 6 mm to 159 mm outside diameters.

GRADES OF COPPER TUBE

BS 2871: Part 1, Table X is light-gauge half-hard tempered tube. It is suitable for installing hot and cold water, gas, heating and sanitation pipework systems above ground.

BS 2871: Part 1, Table Y is light-gauge fully annealed tube, supplied in either coils or straight lengths. It is available with a plastic sheathing applied for protection and identification of con-

Maximum spacing for fixings for copper tube: Table X and Table Z

Nominal diameter of pipe (mm)	Spacing on horizontal run (m)	Spacing on vertical run (m)
15	1.2	1.8
22	1.8	2.4
28	1.8	2.4
35	2.4	3.0
42	2.4	3.0
54	2.7	3.0
67	3.0	3.6
76	3.0	3.6
108	3.0	3.6
133	3.0	3.6
159	3.7	4.2

Maximum spacing for fixings for copper tube: Table Y

Nominal diameter of pipe (mm)	Spacing on horizontal run (m)	Spacing on vertical run (m)
15	1.8	2.0
22	2.4	3.0
28	2.4	3.0
35	2.7	3.0
42	3.0	3.6
54	3.0	3.6
67	3.0	3.6
76	3.6	4.5
108	3.9	4.5

tents (colour of sheath). It is suitable for cold water service pipes, gas and laying under screed floors for heating systems.

BS 2871: Part 1, Table Z is hard-tempered thin-walled tube. It is only suitable for low-pressure systems above ground level. Although it cannot be bent it does have the same outside diameter as pipes in the other tables and therefore may be used with capillary and non-manipulative fittings.

BS 2871: Part 1, Table W is soft-tempered micro- and mini-bore tubing used for heating systems. It is normally supplied in coils for easy application. The wall thickness is slightly less than that of Table Y tube.

Copper pipe bending – Table X copper pipe comes as half-hard tempered and therefore it can be bent cold. The production of accurate bends is best achieved by the use of a 'bending machine', which provides the user with the necessary leverage to pull bends on large-diameter pipes. The alternative to the machine is the use of internal bending springs. These are coiled lengths of tool steel approximately 600 m long which give a flexible support to the pipe bore when the bend is pulled. The skilled user can produce accurate bends, usually limited to 15 mm and 22 mm tubes. Copper pipes can be fully annealed by heating with a blowtorch to make bending easier.
End feed fitting – a type of capillary fitting for copper tube, which contains no solder – the solder is added at the time of jointing from a roll. A comprehensive range of fittings is available, which accommodates all situations. Advantages offered by this type of fitting are that they are relatively cheap and the type of solder used is more flexible, e.g. for heating systems lead–tin solder may be used, and for cold water supply lead-free solder would be used.
Fixings – fixings or brackets are available in many forms for all the materials used to manufacture pipe. Apart from proprietary bracketing systems it is possible to make brackets for special circumstances. The distance between fixings for horizontal and vertical application can be found in various British Standards, particularly BS 6700.

Methods of jointing light-gauge tubes

Tube	Fittings to	BS 864	Other methods		
	Compression Type A	Compression Type B	Capillary (soft solder)	Brazing	Autogenous welding
Annealed copper BS 2871: Table W	Yes	Yes	Yes	Yes	Yes
Half-hard copper BS 2871: Table X	Yes	Yes	Yes	Yes	Yes
Annealed copper BS 2871: Table Y	No	Yes	Yes	Yes	Yes
Half-hard copper BS 2871: Table Y	Yes	Yes	Yes	Yes	Yes
Hard copper BS 2871: Table Z	Yes	No	Yes	No	No

Flux – a substance used to assist the soldering process. Flux usually comes as a paste, which is applied to both pipe end and socket surfaces. There are many proprietary brands of flux based on borax, tallow, resin and zinc chloride. The reason for applying a flux is to prevent oxidization occurring to cleaned surfaces prior to soldering. When heated fluxes become aggressive and tend to dissolve any remaining oxide from the pipe surface. The removal of this coating enables the solder to adhere to the surface of the pipe. Modern fluxes are designed to remove the oxide coating from copper without the need to clean it. Care must be taken to remove any flux residue after jointing to prevent corrosion.

Hard soldering – often referred to as brazing, joint formation is similar to soft soldering joints – a socket is formed in a pipe end and another pipe is inserted into it or alternatively tee joints can be formed using specialist tools. The depth of joint penetration is not as extensive as in the case of soft solder – between 4 mm and 8 mm depending on pipe diameter. The brazing alloy is based on silver, copper and phosphorus or silver, copper, cadmium and zinc. The solder does not require a flux, but it does require great heat to melt the solder. Brazing fittings are available which are similar to end feed fittings.

Identification of pipelines – it is vitally important where there are many pipelines following the same route through a building that it is possible to identify the contents of each pipe. Normally pipework will be insulated to prevent heat loss/gain or to protect the contents from freezing. To this end a system of colour coding has been devised so that by looking at the coding it is possible to know exactly what their contents are. BS 1710 *Identification of Pipelines* sets out the colours for various contents; it specifies the size and location of the coding. For example the base colour for water is green; if that water is potable then an identification code of blue would be applied in strips of 150 mm on top of the green base colour.

Jointing paste – a paste applied to pipe threads prior to the pipe being screwed into a fitting. On wet systems it is used in conjunction with fine strands of hemp. The combination of paste and hemp will fill any spaces left when a joint has been made. Jointing pastes are available for potable water pipes, heating pipes, gas and oil pipes.

Low-carbon steel (LCS) pipe – LCS pipe is manufactured to BS 1387. Pipe is available in three grades, although only two are in common use. The different grades are easily recognized by a colour code painted at one end of the pipe and by the wall thickness.

BS 1387 Light Grade (colour – brown): this is a general-purpose tube suitable for gas, low-pressure heating and in its galvanized form for hot and cold water supply.

BS 1387 Medium Grade (colour – blue): this is a general-grade pipe suitable for hot and cold water supply (galvanized), LPHW heating and gas systems.

BS 1387 Heavy Grade (colour – red): this pipe has the thickest wall and is most suitable for high-pressure heating and steam systems. It is available in sizes from 6 mm to 150 mm diameter. The outside diameter for all three grades is equal. It is therefore the internal diameters that vary, consequently reference is made to the internal nominal bore of a pipe. Jointing methods include threaded, welded and compression.

Low-carbon steel (LCS) pipe bending – low-carbon steel pipe can be bent to form awkward sets and offsets or square bends to assist installation. The easiest method of bending is to use a hydraulic bending machine. An alternative method is the heating up of the pipe until it becomes malleable to pull round. Care should be taken not to flatten the bore of the pipe during this process. The minimum bending radius for LCS pipe is 4 × nominal bore. The bending length (length of pipe to be bent) is the proportion of the circumference subtended by the angle of the bend:

$$\text{Bending length} = \frac{\pi d \times \text{angle}}{360}$$

Example 4.1 Bending radius for a 15 mm nominal bore pipe:

$$\begin{aligned}\text{Bending radius} &= 4 \times \text{nb} \\ &= 4 \times 15 \\ &= \underline{60\ \text{mm}}\end{aligned}$$

Example 4.2 Bending length for a 20 mm nb pipe bent through 90°:

$$\text{Bending length} = \frac{\pi d \times \text{angle}}{360}$$

Maximum spacing for fixings for LCS steel tube

Nominal diameter of pipe (mm)	Spacing on horizontal run (m)	Spacing on vertical run (m)
15	1.8	2.0
20	2.4	3.0
25	2.4	3.0
32	2.7	3.0
40	3.0	3.6
50	3.0	3.6
65	3.0	3.6
80	3.6	4.5
100	3.9	4.5

$$= \frac{3.142 \times 160 \times 90}{360}$$

$$= \underline{126 \text{ mm}}$$

Manipulative compression fitting – a compression fitting that requires work to be done to the end of the pipe to be jointed. This requires the use of special forming tools provided by the manufacturers. They are referred to as Type B fittings. These fittings consist of a body, forming the socket, a nut with an integral friction ring and in certain cases an internal adapter (olive). With the 'Kingley' fitting a special swaging tool is inserted in the end of the pipe which creates a raised bead on the tube before the swaging tool is used. As the nut is tightened onto the body the bead is compressed between the body and the friction ring in the back of the nut. Another type of fitting, 'Securex', requires the end of the pipe to be flared out using a drift and hammer. A conical olive is placed between the end of the fitting and the flared end; the compressive force of the nut holds the assembly together.

Munsen ring – a very robust bracket comprising two halves held together by screws or bolts. The bottom half has a female thread to screw the ring onto its anchor. It is usually used in conjunction with a backplate, which is secured to a wall or floor. Some backplates have a corresponding male thread whilst others have female threads also, so that a spacer piece can complete the bracket. They are manufactured from malleable iron and brass.

Non-manipulative compression fitting – a compression fitting that requires no work to be carried out on the pipe end. These fittings are referred to as Type A fittings. The fitting consists of a body with a thread, which forms a socket, a compression ring or cone and nut, which is tightened on to the body. The nut and ring are slid on to the pipe, which is then inserted into the body. The nut and ring are then pushed up using the appropriate tools. As the nut is tightened a compressive force is applied to the ring which in turn tightens its grip on the surface of the copper.

Pipe thread – threads are cut on pipe ends by a set of four chasing dies, contained in either a hand-threading implement – drophead dies, adjustable stock handles or receder die-head stocks – or a power-threading machine. Also available are hand-held power threaders useful for *in situ* threading work. Threads are cut to BS 21: *British Standard Pipe Thread* (BSPT) either as a tapered or a parallel thread. The length and pitch of thread are described in the standard.

Table of thread engagement lengths

Pipe size (nominal bore)	5	8	10	15	20	25	32	40	50	65	80	100	125	150
Thread engagement Length (mm)	6	8	8	12	14	16	19	19	22	25	28	33	40	40
Total thread Length (mm)	10	12	12	16	18	20	24	24	28	30	35	40	47	47

Plastic clips – either push-in types with open jaws or push-in types with snap-on covers. They are available as single or double clips. Another type commonly used on smaller-diameter pipes is a push-on clip with a hardened nail which can be hammered into most materials.
Plastic pipes – pipes manufactured from PVC (polyvinylchloride), PE (polyethylene) and polybutylene and may be jointed either by solvent welding, push-fit or compression fittings.

Maximum spacing for fixings for plastic tube: MDPE BS 6730

Nominal diameter of pipe (mm)	Spacing on horizontal run (m)	Spacing on vertical run (m)
20	0.5	0.9
25	0.6	1.2
32	0.6	1.2
50	0.8	1.5
63	0.8	1.6

Polybutylene Pipe – a high-grade thermoplastic produced to BS 7291. It has the advantage of being capable of conveying both hot and cold water, but not gas. It is used in the installation of hot and cold water supply and central heating systems. It is produced in long coils and is very flexible. Fittings and valves are available, which are of the push-fit type incorporating an 'O' ring and grab-ring.
Polyethylene pipe – pipe produced in three grades to BS 6572 and BS 6730. The grades refer to the density of the material, e.g. low, medium and high density. These pipes are used for gas and water service pipe and coloured yellow and blue respectively. Polyethylene pipe is jointed by brass compression fittings: a copper insert is pushed into the end of the pipe to take the pressure exerted by the compression ring. Alternatively a push-fit fitting is available which incorporates an 'O' ring and a grab-ring. HDPE – high density polyethylene – pipe is used for mains laying for both gas and water. These pipes are jointed by electrically heated fusion welded joints.

Maximum spacing for fixings for plastic tube: Polybutylene BS 7291

Nominal diameter of pipe (mm)	Spacing on horizontal run (m)	Spacing on vertical run (m)
Up to 16	0.3	0.5
18 to 25	0.5	0.8
28	0.8	1.0
32	0.9	1.2
35	0.9	1.2

Polytetrafluoroethylene (PTFE) – when used as a jointing compound it comes as a tape of various thicknesses. It is used by wrapping it around the thread on a pipe half covering the previous layer until the thread is covered. It should be wrapped around in the opposite direction to tightening the thread into a fitting. General purpose tapes and tapes for gas systems are available.
Push-fit fitting – a range of plastic fittings available for use on copper tube for pressures up to 1.5 bar. The joint seal is effected by a neoprene 'O' ring. A serrated stainless steel grab-ring retains the copper pipe in the fitting.
PVC pipe – pipe manufactured from unplasticized polyvinyl to BS 3505. These pipes are only suitable for cold water supply systems as distribution pipework or mains. PVC pipe has a very smooth bore and therefore offers little resistance to flow. A solvent welding solution joints it: both surfaces of the joint are prepared with a cleaning fluid before the solvent is applied. When the pipe end is pushed into the socket the two surfaces weld together. Several hours need to elapse before full pressure is applied.

Maximum spacing for fixings for plastic tube: UPVC BS: 3505

Nominal diameter of pipe (ins)	Spacing on horizontal run (m)	Spacing on vertical run (m)
¼	0.6	1.1
½	0.7	1.3
¾	0.7	1.4
1	0.8	1.6
1¼	0.9	1.7
1½	1.0	1.9
2	1.1	2.2
3	1.4	2.8
4	1.6	3.1
6	1.9	3.7

Saddle clip – bands of either steel, copper or plastic, which fit over the pipe. They are attached to the substrate by two screws. The main disadvantage of this clip is that it leaves no gap between the pipe and the structure it is fixed to, therefore with water pipes on external walls a pipeboard of wood should be used for protection against freezing.

Skirting board clip – fixings manufactured from malleable iron and brass. Essentially they have two sections, one with holes for fixing it to the wall and a semicircle of the diameter of the pipe, and a half which is fastened to the other half with a screw when the pipe is laid in the clips.

Soft soldering – soft solders are produced to BS 219: 1977 standard from base metals to form alloys. Traditionally soft solders for capillary joints are made from an alloy of lead (40%) and tin (60%). Soft solder can no longer be used on pipework conveying potable water (*Model Water Byelaws* 1986). For potable water systems (hot and cold) lead-free solder is now required. This is an alloy of tin (87%) and silver (3%) also produced to BS 219 (amended December 1987).

Solder ring fitting – a capillary fitting with an integral lead-free solder ring within the socket. The ring contains sufficient solder to complete the joint and therefore no further solder should be applied. There are many types and configurations of fittings available in all sizes up to 65 mm diameter. Fittings above this size have special jointing techniques.

Thermoplastics – a category of plastics that are softened and manipulated by heat and pressure without altering the chemical structure of the material.

Threaded fittings – threaded pipe fittings are available from many manufacturers. They are produced from malleable iron, low-carbon steel and brass. The threads are to BS 21 (BSPT) and are tapped as parallel or tapered threads. Fittings are available as sockets, elbows, bends, tees, unions, bushes, nipples and flanges. Malleable iron fittings usually have a bead cast on the opening of the fitting for additional strength.

Valves – pipeline fittings or components that are used to control the flow of fluid in the pipe. They may operate automatically or by manual means. Valves are manufactured from bronze or brass. Obturating valves such as globe valves, stop valves, and gate valves close slowly and therefore do not normally create pressure surges in the system, which could cause water hammer to occur. Float-operated valves (ball valves) are used to maintain the water level in storage and flushing cisterns automatically. Service valves are inline valves of the quick shut-off type. A spherical valve with a hole drilled through it, held in PTFE seals, opens and closes by turning it through 90°. It is turned by a flat-bladed screwdriver. Service valves should be installed so as to isolate float-operated valves.

Welded joints – a specialist joint involving a high level of skill. Welding techniques include oxyacetylene gas welding and manual metal arc welding. Pipe welding is an autogenous process. Weldon fittings are available or alternatively a joint may be fashioned from the pipe.

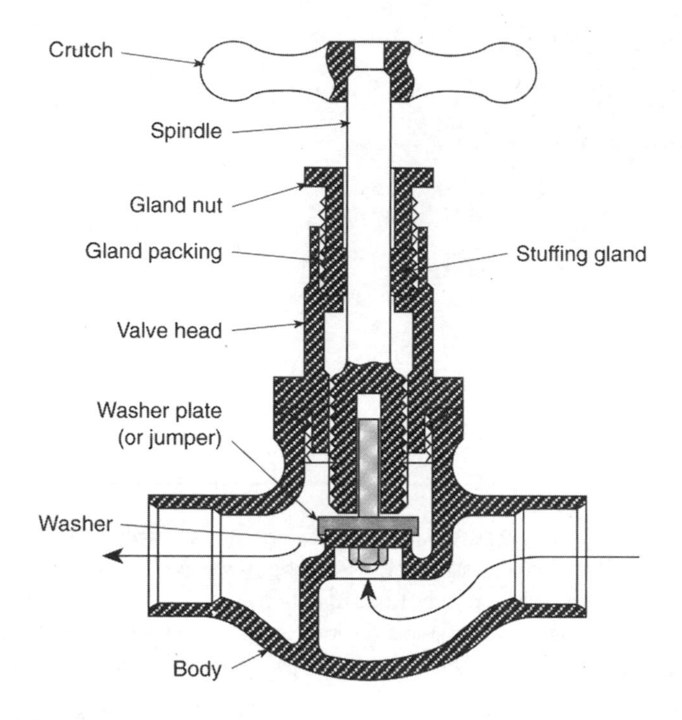

Fig. 4.1 Stop tap to BS 5433

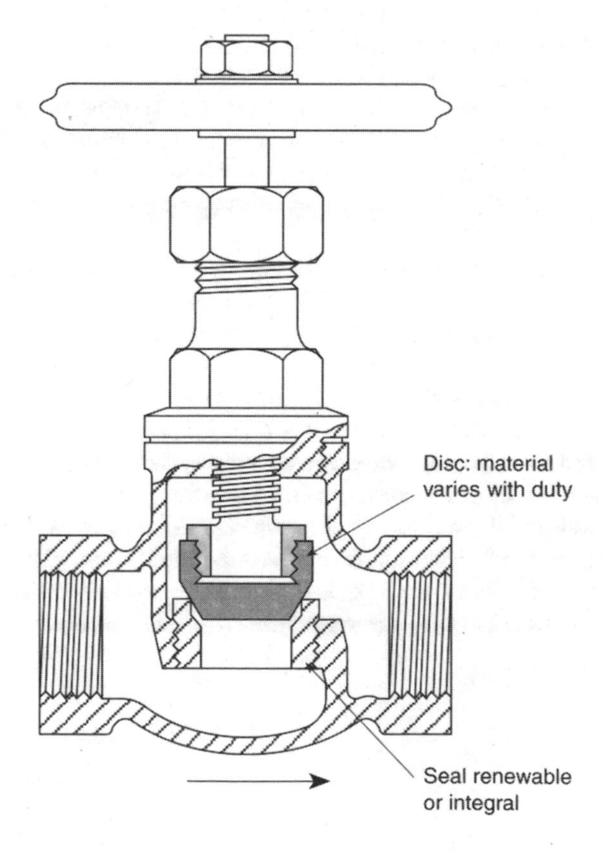

Fig. 4.2 Globe valve

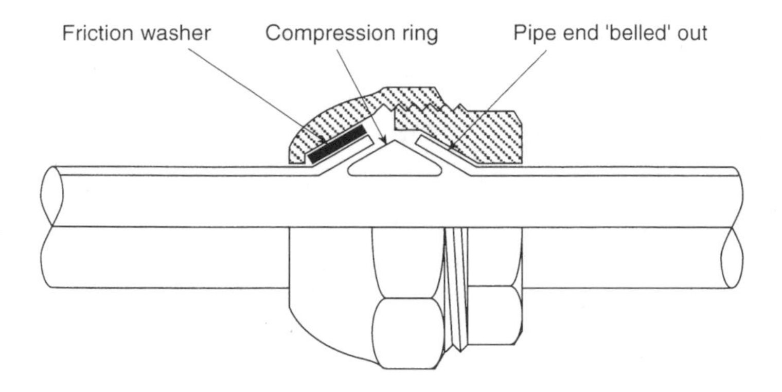

Fig. 4.3 Manipulative compression fitting

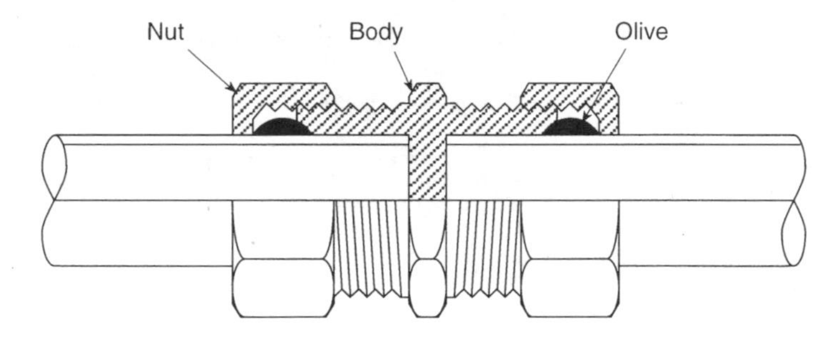

Fig. 4.4 Non-manipulative compression fitting

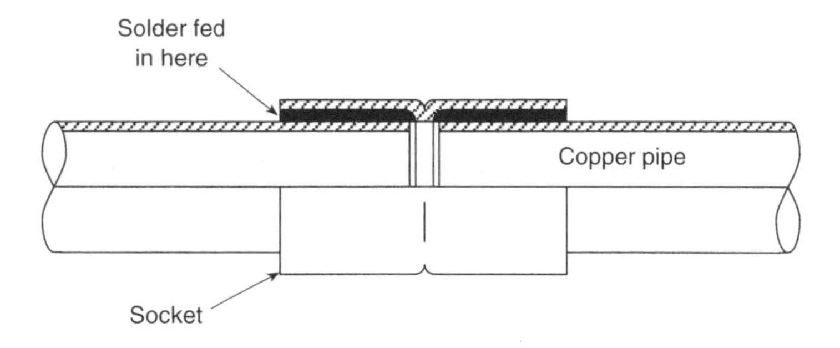

Fig. 4.5 Jointing method for light-gauge copper pipe. End-feed capillary soft solder

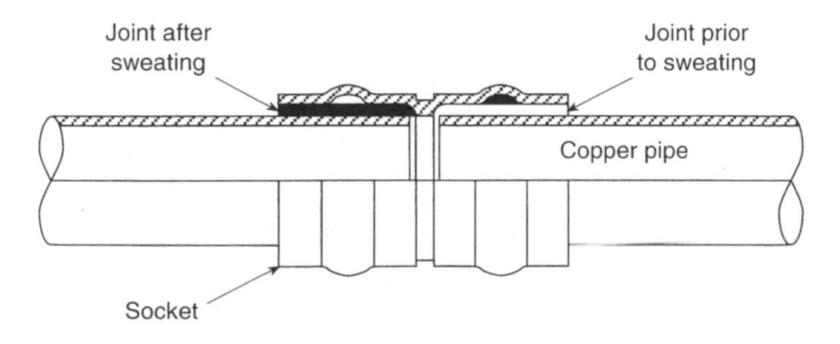

Fig. 4.6 Integral solder ring capillary soft solder

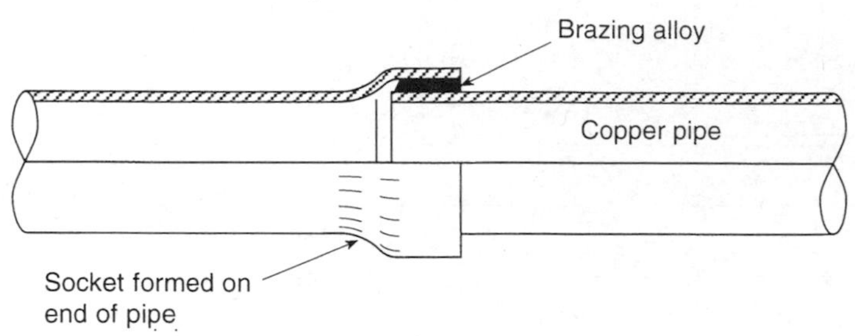

Fig. 4.7 Brazed capillary – hard solder

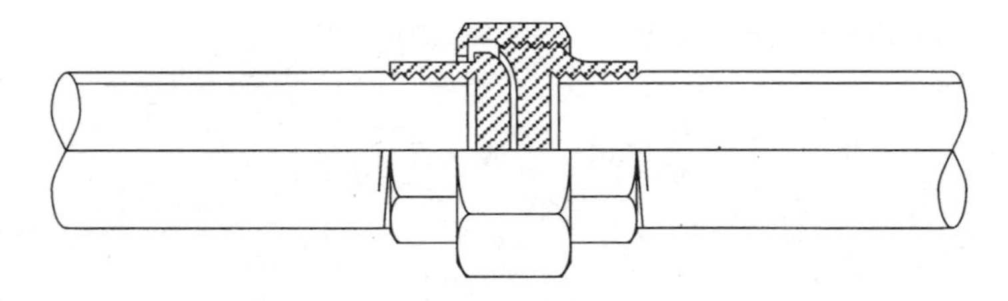

Fig. 4.8 Union fitting for low-carbon steel pipe

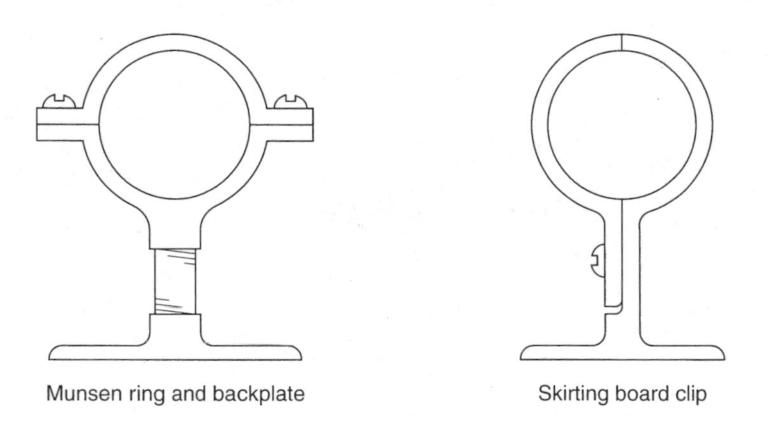

Fig. 4.9 Pipe fixings

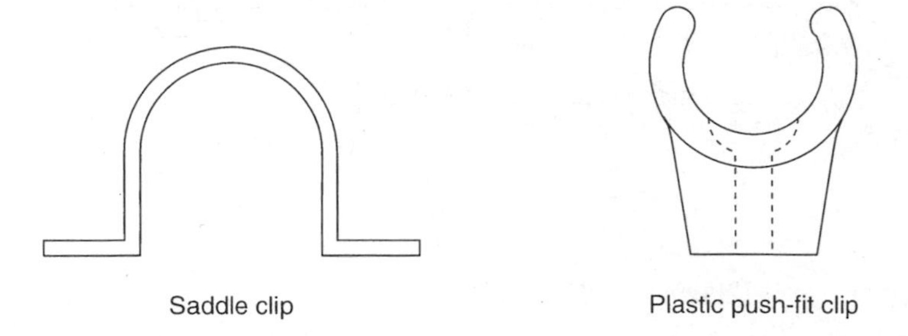

Fig. 4.10 Pipe fixings

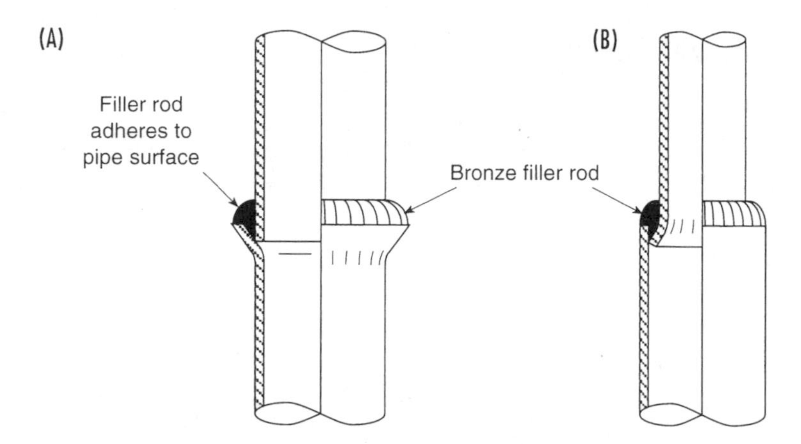

Fig. 4.11 (A) Bronze welded bell butt joint; (B) Bronze welded reducing bell butt joint.

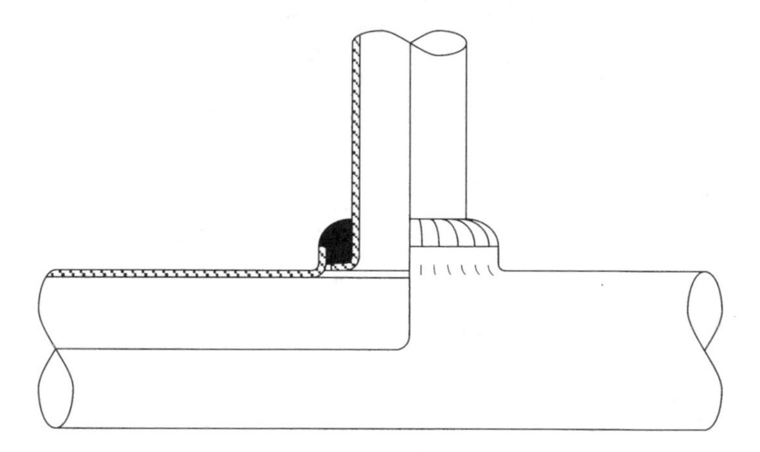

Fig. 4.12 Bronze welded branch joint

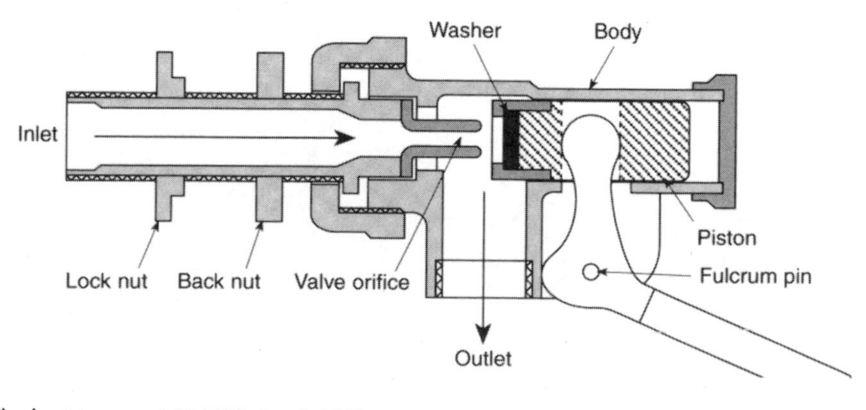

Fig. 4.13 Ball valve (piston type) BS 1212: Part 1: 1953

Now try and answer these questions:

1. The grade of copper tube that should not be bent is
   - a Table X
   - b Table Y
   - c Table Z
   - d Table W.

2. Copper tube is manufactured to ______ Part 1 for light gauge tubing.
   - a BS 2871
   - b BS 6700
   - c BS 6644
   - d BS 1387.

3. The alloy used in the bronze welding process is produced from
   - a copper, tin and lead
   - b brass, tin and lead
   - c brass, zinc and nickel
   - d copper, zinc and nickel.

4. A type of joint on copper tube where molten solder is drawn into the joint is referred to as a
   - a thread joint
   - b solvent joint
   - c capillary joint
   - d compression joint.

5. The grade of copper pipe most suited for cold water service pipes is
   - a Table X
   - b Table Y
   - c Table Z
   - d Table W.

6. The maximum spacing of horizontal fixings on 28 mm Table X copper pipe is
   - a 1.8 m
   - b 2.0 m
   - c 2.8 m
   - d 3.0 m.

7. The maximum spacing of vertical fixing on 54 mm Table Y copper pipe is
   - a 2.0 m
   - b 2.6 m
   - c 3.0 m
   - d 3.6 m.

8. The bending radius of a bend on 15 mm diameter LCS pipe is
   - a 90 mm
   - b 120 mm
   - c 60 mm
   - d 45 mm.

9. The bending length of a 90° bend 20 mm diameter LCS pipe is
   - a 88 mm
   - b 126 mm
   - c 200 mm
   - d 232 mm.

10. The contents of pipelines are identified by
    - a a sign
    - b a bar code
    - c a colour code
    - d pipe type.

11. The depth of joints for copper pipe when the brazing technique is used should be
    - a 4 to 8 mm
    - b 15 to 22 mm
    - c 2 to 6 mm
    - d 22 to 28 mm.

12. The jointing method referred to as 'brazing' is also known as
    - a bronze welding
    - b autogenous welding
    - c soft soldering
    - d hard soldering.

13. One source of information regarding the maximum spacing of pipe fixings could be

a BS 6644
b BS 6700
c BS 1192
d BS 1553.

14. The noble metal found in brazing alloys is

a gold
b platinum
c silver
d duridium.

15. A flux paste is applied to both pipe end and the internal socket surfaces on copper pipe to prevent ____________ prior to soldering.

a tinning
b fusing
c oxidization
d fluctuation.

16. Which one of the following is used in the production of fluxes?

a zinc chloride
b detergent
c potassium
d potash alum.

17. The most suitable hand tool for bending light gauge copper pipes would be

a lump hammer
b internal spring
c return spring
d pipe gauge.

18. The production of accurate bends on copper pipe is best achieved by using a

a manual
b internal spring
c metric ruler
d bending machine.

19. On wet systems constructed with screwed LCS pipe and fittings, jointing paste is used in conjunction with

a flux
b threading compound
c hemp
d PTFE.

20. Low-carbon steel pipe is manufactured to

a BS 2871
b BS 1387
c BS 216
d BS 476.

21. Low-carbon steel pipe can only be used for hot and cold water systems if it is firstly

a sterilized
b purged
c painted
d galvanized.

22. For the installation of high-pressure systems the grade of LCS pipe used must be

a heavy
b medium
c light
d half weight.

23. The maximum spacing for horizontal pipe runs on 32 mm diameter LCS pipe is

a 3.0 m
b 3.6 m
c 2.4 m
d 2.7 m.

24. A compression fitting which requires the end of the copper pipe to be worked is referred to as a

a non-manipulative fitting
b primofit fitting
c manipulative fitting
d victaulic fitting.

25. Non-manipulative compression fittings are referred to as __________ fittings.

a Type A
b Type B
c Type C
d Type D.

26. Pipe threads used on LCS pipe are cut to __________. *British Standard Pipe Thread* (BSPT).

a BS 21
b BS 41
c BS 88
d BS 19.

27. The total thread length for 50 mm LCS pipe is

a 32 mm
b 15 mm
c 28 mm
d 40 mm.

28. HDPE pipe used for gas and water mains is jointed by

a solvent cement
b fusion welding
c capillary attraction
d push-fit fittings.

29. Polyethylene pipe is manufactured to both

a BS 1192 and BS 1553
b BS 1387 and BS 2871
c BS 6644 and BS 5440
d BS 6572 and BS 6730.

30. Polybutylene pipe is manufactured to

a BS 7291
b BS 1710
c BS 1387
d BS 2871.

31. Fittings for use on polybutylene pipe are push-fit and incorporate a

a 'D' ring and grab-ring
b friction washer
c 'O' ring and grab-ring
d compression ring.

32. Soft solder produced to BS 219: 1977 is traditionally an alloy of

a lead and tin
b copper and tin
c lead and silver
d zinc and tin.

33. Which document prohibits the use of lead/tin solders on potable water systems?

a *Gas Safety Regulations 1974*
b *Model Water Bylaws 1986*
c *IEE 16th Edition Regulations*
d BS 7276.

34. Soft solder which is permissible on potable water systems is an alloy of

a silver and tin
b lead and tin
c copper and tin
d zinc and tin.

35. The bending length for a 45° bend on 15 mm LCS pipe would be

a 100 mm
b 147 mm
c 37 mm
d 47 mm.

36. The bending length of a 90° bend on 25 mm LCS pipe is

a 250 mm
b 90 mm
c 157 mm
d 125 mm.

37. Plastics that are capable of being softened and manipulated without altering their chemical structure are classified as

a thermosetting
b thermoplastics
c plasticised
d bakelite.

38. Pipeline components that are used to control the flow of fluids are referred to as

a restrictors
b compressors
c valves
d impeders.

39. The device that is used to maintain the water level in a cistern is known as a

a regulating valve
b service valve
c gate valve
d float-operated valve.

40. A plastic push-fit fitting suitable for use on copper pipe is limited to a system pressure of

a 1.5 bar
b 0.5 bar
c 2.5 bar
d 1.5 bar.

41. The identification of pipelines is described in

a BS 1192
b BS 1553
c BS 1710
d BS 1387.

42. A type of pipe bracket which is very robust is known as a

a skirting board clip
b Munsen ring
c saddle clip
d plastic clip.

43. The maximum spacing of horizontal fixing for 18 mm polybutylene pipe is

a 1.0 m
b 0.6 m
c 1.2 m
d 0.5 m.

44. The welding of LCS pipe is a type of welding process known as

a bronze welding
b autogenous welding
c soft soldering
d hard soldering.

45. Malleable iron pipe fittings usually have ________ cast on to the opening of the fitting.

a a bead
b an arrow
c a letter
d a flange.

Now check your answers from the grid.

Q 1; c	Q 10; c	Q 19; c	Q 28; b	Q 37; b
Q 2; a	Q 11; a	Q 20; b	Q 29; d	Q 38; c
Q 3; d	Q 12; d	Q 21; d	Q 30; a	Q 39; d
Q 4; c	Q 13; b	Q 22; a	Q 31; c	Q 40; b
Q 5; b	Q 14; c	Q 23; d	Q 32; a	Q 41; c
Q 6; a	Q 15; c	Q 24; c	Q 33; b	Q 42; b
Q 7; d	Q 16; a	Q 25; a	Q 34; a	Q 43; d
Q 8; c	Q 17; b	Q 26; a	Q 35; d	Q 44; b
Q 9; b	Q 18; d	Q 27; c	Q 36; c	Q 45; a

5

GAS SUPPLY

To tackle this chapter you will need to know:

- Gas Safety (Installation & Use) Regulations 1994;
- gas distribution systems;
- internal gas installation;
- control of pressure in gas systems;
- ventilation requirements for safe combustion;
- testing of gas installations;
- testing of flues for gas appliances;
- gas rating of gas appliances.

GLOSSARY OF TERMS

Anaconda – a flexible connection which is inserted between the service control valve and the service governor. It enables the meter to be easily removed and prevents unnecessary stress from two fixed connections. Manufactured from stainless steel, it should never be bent less than 1.5 times its radius to prevent failure.

Balanced flued appliance – an appliance which has a flue outlet and air inlet which are at the same point on the external surface of the property wall, hence the term 'balanced flue'. This type of appliance requires no additional ventilation unless installed in a compartment.

Closure plate – a plate of aluminium sheet which is designed to seal the fireplace opening prior to the installation of a gas fire. The closure plate is sealed into place by special sealing tape. The plate has two holes cut into it; one for the flue spigot of the appliance and one to act as a balancing hole for the flue draught.

Compartment – an enclosure within a room which contains a gas appliance and requires special consideration when assessing the ventilation rates for appliances.

Complete combustion – the burning of gas with sufficient ventilation provided to ensure there is ample air to produce carbon dioxide and water vapour as products of combustion. To achieve this, ventilation should be 4.5 cm^2 of free air space for each kilowatt in excess of 7 kW of heat input; but check the manufacturers' instructions.

Constant pressure governor – is a device used to set the gas pressure flowing to a gas appliance at a pre-set value for the purpose of providing the stated heat input. It works on the principle of gas acting upon the surface of a plastic diaphragm which is countered by a spring acting upon the other side. The spring tension determines the downstream gas pressure.

Continuity bonding – With regard to the *IEE* 16th *Edition Regulations* and the *Gas Safety (Installation & Use) Regulations 1998* all gas installations must be connected to the earth protection system of an electrical installation. The connection must be within 600 mm of the outlet of the gas meter.

Distribution pipework – the system of pipework which connects the gas meter to all the connected appliances. It is now normally installed with copper tubing utilizing capillary soldered joints. The diameter of the pipes is dependent upon the gas flow rate of the appliances connected.

Flueless appliance – an appliance which possess no flue and therefore discharges the products of combustion into the room where it is installed, e.g. a cooker or single-point water heater. This type of appliance should be in a room with a door or window opening directly to the outside air.

Gas appliance – is a purpose-built device constructed to burn gas to heat water for domestic use, heat wet central heating systems, provide space heating or for cooking food.

Gas cock – a control device intended for the isolation of a section of pipe or gas appliance for the purpose of removal or servicing. It is normally a brass tapered plug-type cock which turns off the supply with a 90° turn action.

Gas main – a system of distribution pipework supplying natural gas to the customers of the suppliers. The system is the property of Transco which is the transportation side of British Gas PLC. The gas which is used by the consumers is not necessarily the property of British Gas, but may be supplied by an independent supplier. In modern times gas mains are laid using polyethylene high-density pipe (PE) in preference to cast iron, which is the traditional material.

Gas meter – a measuring device which is the property of the gas supplier and is used as a basis for the calculation of the user's gas bill. A typical size of meter for a domestic supply would be either a U6 or E6.
Gas rate – the amount of gas consumed by an appliance within one hour and is measured in m^3/hr (cubic metres per hour). The gas rate is directly related to the maximum heat input. Gas rating is an important part of commissioning and rectifying faulty appliances.
Manometer (U gauge) – a very simple, yet very effective, measuring instrument capable of measuring among other fluids natural gas (or any other gas). Its structure is simply a U-shaped tube filled with water, with a calibrated scale attached. The calibrations represent either one or two mbar. One end of the U is attached to the gas test nipple and the other is open to the atmosphere. The difference in water levels within both legs of the gauge is indicative of the prevailing pressure in the pipework in excess of atmospheric pressure.
Meter union – a special fitting which is designed to connect the distribution pipework to the gas meter by means of a compression joint using a rubber ring seal to effect the joint. It can be to either 22 mm or 28 mm copper pipe.
Open flued appliance – an appliance which is connected to a chimney or purpose-built flue. It would normally have a draught diverter at the end of the primary flue pipe. This appliance gets its combustion air from within the space it is installed in. Typical appliances might be a gas boiler or a multi-point water heater.
Pipe clips – devices for fixing pipework to a wall or under-floor surface to support the pipework and therefore preventing it from sagging and fracturing. They must be spaced within the constraints of the appropriate British Standard.
Pipe sleeve – where a gas pipe passes through a wall or a floor structure it must be encased in a pipe sleeve to prevent any potential gas leeks within the structure from infiltrating into the fabric of the building. The sleeve is usually a piece of similar pipe material, but of one or two diameters above the pipe in question.
Purge – with reference to domestic systems the term to 'purge' means the evacuation of the air/air–gas mixture from the pipework after installation or rectification of pipework systems. The procedure for domestic systems is basically to open the gas taps of the furthest appliance from the meter and turn on the service control valve until five times the badge capacity of the meter has passed through the meter (for a U6 meter this is 0.335 ft^3).
Restrictor elbow – a control device normally associated with gas fires and a means of isolation and disconnection from the distribution pipework. It is to be found at the side of the gas fire where the pipe comes up through the floor.
Service control valve – a lever action plug of a ball-type valve which is fitted as close to the inlet to a property as is practicable. Alternatively it will be sited within a wall-mounted meter box externally. The service control valve is also the emergency shut-off valve and therefore the lever should fall to close the valve.
Service governor – a compensated constant-pressure governor which reduces the mains pressure (usually 30 mbar) to a typical working pressure of 21 mbar for a domestic supply.
Service pipe – an underground pipe which connects the property to the gas main. The modern trend is to lay the service pipe in PE and terminate in the building with galvanized steel tube through the walls of the property, or in the case of an external meter box to rise from the ground to the meter box in steel pipe.
Smoke match – as the name suggests this is a special type of match which generates white smoke when ignited. The match is used to test an appliance for 'spillage' of the product of combustion entering the room where the appliance is situated.
Smoke pellet – a smoke generator for the important testing of a chimney or flue for suitability for connecting a gas appliance to it.
Standing pressure – operating pressure recorded by a manometer. It is a reading of the gas pressure within a system when all the appliances are turned off and no gas is flowing. A normal domestic gas system would register a standing pressure of 21 mbar. Should there be a leak in the system the manometer reading will invariably show a pressure drop.
Test dial – in the case of the U6 meter this dial measures 1 ft^3 per revolution. The purpose of the dial is in the calculation of the gas rating of appliances. In the case of the E6 meter the measurement is 0.001 m^3.
Test nipple – an adapter tapped into the outlet side of the gas meter, intended for the purpose of testing the internal gas pipework system for soundness. It is a brass component which has a pap capable of accepting the rubber tube from the U gauge and is sealed after use by a screwed bung.
Ventilation rate – the amount of free air ventilation into a room containing one or more gas appliances. The amount of permanent ventilation is dependent upon the type of appliances and the maximum heat input of the appliances plus the room volume.
Working pressure – the gas pressure measured at the upstream side of an appliance governor when the appliance is operating at maximum heat input. The pressure recorded must not read below one mbar of working pressure at the meter. Any pressure reading greater than one mbar indicates at best a blockage or at worst the undersizing of the pipework feeding the appliance.

Gas rating exercise

The gas rate of an appliance is the amount of gas burnt by the burner in 1 hour. Because **pressure is inversely proportional to volume** (Boyles law), the pressure at the burner, i.e. the burner pressure, will determine the amount of gas being burnt. Simply the greater the pressure the greater the amount of gas burn, the more gas you burn the greater the gas rate (more gas, more heat).

All fuels have what is called a **calorific value** (CV), that is to say an amount of heat potential within a known amount of fuel. For instance we talk of the calorific value of gas as the number of megajoules per cubic metre (MJ/m^3). Therefore if you increase the amount of gas you burn it follows you increase the heat input to the appliance.

For natural gas the average calorific value is 38 MJ/m^3. This information is on all gas bills or is freely available from the supplier (their name is to be found on the gas meter).

All appliance manufacturers are required to state the gas rate for an appliance for given heat inputs and outputs. You will find this information on the appliance 'badge'.

Anyone who installs, services or commissions an appliance is required by law to gas rate the appliance and check against the quoted value.

The method of gas rating an appliance is to

Firstly run the appliance for 5 to 10 minutes

Secondly time the test dial of the meter through one revolution

Thirdly note the time taken, then calculate the gas rate.

Gas rate tables are available in publications or a simple calculation will give you the desired solution:

$$\text{Gas rate} = \frac{1 \text{ Rev} \times 3600 \times 0.028}{\text{seconds}}$$

For instance if 1 revolution of the dial takes 50 seconds the gas rate will be

$$\text{Gas rate} = \frac{1 \text{ rev} \times 3600 \times 0.028}{\text{seconds}}$$

$$= \frac{100.8}{50}$$

$$= \underline{2.016 \text{ m}^3/\text{h}}$$

1 revolution is equal to 1 cubic foot on U6 (domestic) meters.

The number 0.028 is a factor which converts cubic feet to cubic metres.

Now you can try a selection of gas rate calculations for yourself. Please complete the following examples.

1. The test dial of a gas meter serving a gas fire was observed to complete one revolution in 80 seconds. Calculate the gas rate of the fire.

($1.26\ m^3/h$)

2. A multipoint water heater working at full load caused the test dial of the gas meter to complete one revolution in 25 seconds. What is the gas rate of the appliance?

($4.032\ m^3/h$)

3. If the test dial of the gas meter feeding a heating system boiler takes 32 seconds to complete one revolution, what is the gas rate of the boiler?

($3.15\ m^3/h$)

4. An appliance operating at full capacity causes the test dial to revolve in 2 minutes and 10 seconds. Calculate the gas rate of the appliance.

($0.76\ m^3/h$)

It might be a useful exercise to now construct a table which you can use to read off gas rates conveniently when you are out on site working, without the need to calculate the gas rate.

Complete the table below by calculating the gas rates. To assist you and make your task easier you may have already noticed that if you multiply the two numbers on the top line, that is 3600 × 0.028 you get 100.8. These are constant numbers so therefore all that is necessary is for you to divide 100.8 by the time in seconds.

Time in seconds for 1 revolution of test dial	Gas rate in m^3/h	Time in seconds for 1 revolution of test dial	Gas rate in m^3/h
20		41	
21		42	
22		43	
23		44	
24		45	
25		46	
26		47	
27		48	
28		49	
29		50	
30		51	
31		52	
32		53	
33		54	
34		55	
35		56	
36		57	
37		58	
38		59	
39		60	
40		61	

Heat input

All fuel-burning appliances, and gas appliances are no exception, have quoted **heat inputs** and **outputs**. That is to say that the fuel burnt will liberate a set amount of heat energy during the combustion process. Because the products of combustion have to be removed to the outside atmosphere there is obviously going to be a heat loss through the flue products. Therefore we have a heat output. This is the useful heat from the appliance.

When working on gas appliances, whether it be installing, servicing or commissioning, the operative is obliged, under the *Gas Safety (Installation and Use) Regulations 1998*, to gas rate and verify the heat input of that appliance before he leaves it. Fortunately we can utilize the same calculation that we have just used to calculate the gas rate by introducing the **calorific value** of the fuel.

As previously stated the average calorific value of natural gas is 38 MJ/m^3 of gas. Another fact that is useful to know is that **1 kW is equivalent to 3.6 MJ** of heat energy.

With these two factors in mind let us see how we can manipulate the gas rate formula to give heat input values.

Firstly

$$\text{Gas rate} = \frac{1 \text{ Rev} \times 3600 \times 0.028}{\text{seconds/rev}}$$ and gives us m^3/h

Secondly — if we multiply the gas rate by the calorific value we get the heat energy of the fuel per hour being burnt

Thirdly — to convert the heat energy to kilowatts we must divide the value of the heat energy by the megajoules per kilowatt

Finally

$$\text{Heat input} = \frac{1 \text{ Rev} \times 3600 \times 0.028 \times \text{CV}}{\text{seconds/rev} \times 3.6}$$

Let us now apply this new development to our original gas rate example.

An appliance burning gas causes the test dial of the meter to complete one revolution in 50 seconds. Calculate the heat input of the appliance.

$$\text{Heat input} = \frac{1 \text{ Rev} \times 3600 \times 0.028 \times \text{CV}}{\text{seconds/rev} \times 3.6}$$

$$= \frac{1 \times 3600 \times 0.028 \times 38}{50 \times 3.6}$$

$$= \frac{3830.4}{180} = 21.28 \text{ kW}$$

Now it's your turn again. Take the previous four examples of gas rates and calculate the heat inputs for those four appliances:

1. ______________________________

____

____

____

____

____

(13.3 kW)

2. ____

____

____

____

____

____

(42.56 kW)

3. ____

____

____

____

____

____

(33.25 kW)

4. ____

____

____

____

____

____

(8.02 kW)

Now you have completed the examples you should be able to accomplish several simple calculations with regard to heat and gas flow.

Now try and answer these questions

1. The gas main system of underground pipework which is used by gas suppliers to deliver gas to their customers is maintained by a division of British Gas PLC known as
   - **a** Gas Force
   - **b** Natgas
   - **c** Transco
   - **d** Maingas.

2. The modern pipe material being used to lay new or replacement gas main underground is
   - **a** polyethylene
   - **b** polypropylene
   - **c** low-carbon steel
   - **d** cast iron.

3. The small-diameter pipe carrying the gas from the gas main to a property is known as the
   - **a** distribution pipe
   - **b** conduit
   - **c** service pipe
   - **d** communication pipe.

4. Before entering the building or rising to an external meter box the pipe conveying the gas to the property changes material for added strength to
   - **a** high-density PE
   - **b** cast iron
   - **c** lead
   - **d** galvanized low-carbon steel.

5. The supply to the building is controlled by a device referred to as the
   - **a** gas cock
   - **b** service control valve
   - **c** main gas cock
   - **d** gate valve.

6. The lever of the control device in question 5 should for emergency purposes
   - **a** be left at the side of the meter
   - **b** fall to the closed position
   - **c** be pulled upwards to close the valve
   - **d** be locked in the open position.

7. A component known as an 'anaconda' is fitted between the service control valve and the service governor. The reason for this is to provide
   - **a** an easy way to join the meter to the supply
   - **b** a means of expansion for the gas pipes
   - **c** a device for absorbing vibrations
   - **d** a flexible connection to facilitate meter removal.

8. The gas within the anaconda will be at same pressure as the
   - **a** internal pipework
   - **b** gas meter
   - **c** gas mains
   - **d** transmission mains.

9. The 'service governor' is a device which is mounted on the inlet side of a gas meter. Its function is to
   - **a** measure the amount of gas used
   - **b** join the anaconda to the meter
   - **c** calculate the gas rate of an appliance
   - **d** reduce mains pressure to working pressure.

10. A 'service governor' is a type of governor referred to as a
    - **a** compensated constant-pressure governor
    - **b** constant-pressure governor
    - **c** balanced-pressure governor
    - **d** weight-loaded governor.

11. The size of a domestic gas meter is typically either a

a A6 or G6
b U6 or E6
c U16 or E16
d B4 or C4.

12. The gas meter installed in a dwelling remains the property of the

a householder
b Local Authority
c landlord
d gas supplier.

13. On the outlet side of the gas meter is a special tapping which is referred to as a

a bleed nipple
b spare tapping
c test nipple
d vent nipple.

14. The special tapping on the outlet side of the gas meter is a means of attaching an instrument for the purpose of

a testing the quality of the gas
b testing for soundness of the pipework
c measuring the length of pipework
d estimating the gas bill.

15. The instrument used to test gas pipework for soundness is called a

a manometer
b bourdon gauge
c gasometer
d gascoseeker

16. The instrument referred to in Question 15 consists of a structure involving a

a diaphragm connected to a pointer
b expanding and contracting bellows
c clear 'U' shaped tube containing water
d system of connecting levers.

17. The standing pressure of a gas system is measured when

a all the gas appliances are operating fully turned on
b half the gas appliances are turned on
c the supplier connects the supply
d the gas is turned on and all gas appliances are turned off.

18. The standing pressure of a domestic gas system is set by the gas supplier at:

a 26 mbar
b 21 mbar
c 30 mbar
d 31 mbar.

19. The working pressure of a gas system is measured at the gas appliance when

a the appliance is turned off
b the appliance is working at half heat input
c the appliance is disconnected
d the appliance is operating at maximum heat input.

20. If the working pressure at an appliance is greater than 1 mbar it could be indicative of

a inadequate pipe sizing
b high-density gas
c a faulty appliance
d lack of ventilation.

21. On completion of a gas pipework installation the system should be purged. The reason for purging is to

a remove flux residue
b prevent back pressure
c check for leakage
d remove air and gas/air mixtures prior the use.

22. In a domestic gas installation the approximate amount of gas allowed to pass through the meter to effect a purge is

a 0.335 m^3
b 0.335 ft^3
c 0.017 ft^3
d 0.017 m^3.

23. The test dial on the panel of the gas meter is used to determine either of the following

a gas bill or gas rate
b gas bill or purge volume
c leakage rate or purge volume
d gas rate or purge volume.

24. All gas appliances have a stated maximum heat input declared by the manufacturer. This relates to the amount of gas consumed and is known as the

a consumption rate
b gas value
c gas rate
d calorific value.

25. The gas rating of an appliance is an important part of the safe _________ of the appliance.

a de-commissioning
b commissioning
c pricing
d selection.

26. The internal system of pipes connecting the meter to all the appliances within a property if known as the

a distribution pipework
b internal pipework
c service pipe
d rising main.

27. The selection of pipe sizes for gas installations is determined by the sum of the _____ of the appliances connected to it.

a burners
b heat outputs
c gas rates
d gas pressures.

28. The continuity bonding connection clamp must be made within _______ of the outlet of the meter.

a 1000 mm
b 600 mm
c 750 mm
d 1200 mm.

29. If gas pipework has to pass through walls and solid floor structures it must be encased in a continuous

a sand and cement mortar
b mastic
c concrete
d pipe sleeve.

30. The device intended to provide isolation for the purpose of servicing appliances is known as a

a ball cock
b gas cock
c gate valve
d stop valve.

31. The operating pressure of a gas appliance is set and maintained by a device known as a

a service governor
b pressure reducing valve
c relay valve
d constant pressure governor.

32. The device referred to in Question 31 relies on gas acting upon a

a plastic diaphragm countered by a spring
b swing gate type valve
c spring loaded valve
d pressure build-up in a pilot tube.

33. Where a gas fire is connected to the gas distribution pipe a control device referred to as a __________ is commonly used.

a gas tap
b knuckle valve
c restrictor elbow
d gate valve.

34. The amount of free air ventilation provided for a gas appliance is based on the maximum heat input and is referred to as the

a ventilation brick
b ventilation rate
c air change rate
d air brick.

35. The amount of free air ventilation provided for a gas appliance having a heat input in excess of 7 kW must be

a $10\,cm^2$ per kW
b $5.4\,cm^2$ per kW
c $4.5\,cm^2$ per kW
d $100\,cm^2$ per kW.

36. A gas boiler with a maximum heat input of 17 kW must be provided with __________ cm^2 of free air space.

a $4.5\,cm^2$
b $60\,cm^2$
c $45\,cm^2$
d $76.5\,cm^2$.

37. Where an appliance is suspected of having a faulty flue, a means of testing for spillage might be the use of a

a smoke pellet
b manometer
c swan vesta match
d smoke match.

38. When testing a chimney for its suitability for connecting a gas appliance to it, a __________ is placed at the chimney base and ignited.

a smoke pellet
b smoke match
c paraffin soaked rag
d sheet of newspaper.

39. A gas appliance which is not connected to a flue or chimney and issues its products of combustion into the space in which it is installed is known as

a a water boiler
b an open-flued appliance
c a flueless appliance
d balanced-flued appliance.

40. The type of appliance referred to in Question 39 should be installed in a space which

a is referred to as a compartment
b has a door or window opening to outside air
c is used as a bedroom
d is used as a bathroom.

41. An appliance which is connected to a chimney or flue and takes combustion air from within the space in which it is situated is referred to as

a an open-flued appliance
b a balanced-flued appliance
c a flueless appliance
d a cabinet heater.

42. An appliance similar to the type referred to in Question 41 will have a device called a __________ connecting the primary and secondary flues.

a flue terminal
b flue guard
c closure plate
d draught diverter.

43. A type of gas appliance which receives its combustion air adjacent to the flue outlet on an exterior wall surface is known as

a a flueless appliance
b a balanced-flued appliance
c an open-flued appliance
d a combi-boiler.

44. A balanced-flued appliance requires no free air provided ventilation unless it is installed within a

a bedroom
b bathroom
c caravan
d compartment.

45. The term 'compartment' in relationship to gas installations is given to

a accommodation for people
b a section of the combustion chamber
c an enclosure specifically for gas appliances
d a room specifically for installing gas appliances.

Now check your answers from the grid.

Q 1; c	Q 10; a	Q 19; d	Q 28; b	Q 37; d
Q 2; a	Q 11; b	Q 20; a	Q 29; d	Q 38; a
Q 3; c	Q 12; d	Q 21; d	Q 30; b	Q 39; c
Q 4; d	Q 13; c	Q 22; b	Q 31; d	Q 40; b
Q 5; b	Q 14; b	Q 23; d	Q 32; a	Q 41; a
Q 6; b	Q 15; a	Q 24; c	Q 33; c	Q 42; d
Q 7; d	Q 16; c	Q 25; b	Q 34; b	Q 43; b
Q 8; c	Q 17; d	Q 26; a	Q 35; c	Q 44; d
Q 9; d	Q 18; b	Q 27; c	Q 36; c	Q 45; c

6

OIL SUPPLY

To tackle this chapter you will need to know:

- types of oil supply pipe systems;
- types of oil burners and their components;
- methods of oil storage;
- types of oil pipework components;
- the combustion and consumption of oil.

GLOSSARY OF TERMS

Air diffuser – often referred to as a 'swirl ring', this is a specially designed ring with slots or cuts around its circumference. It creates turbulence in the flow of combustion air allowing efficient mixing of the atomized oil and air. It is situated in a position in front of the burner nozzle; the actual position is a vital factor in flame stabilization.
Air fan – a means of supplying combustion air to the burner under pressure. The air is pressurized by a centrifugal fan impeller mounted directly on the motor shaft. The shaft incorporates a shearing device to protect the motor in the event of the impeller jamming. A damper on the inlet to the fan adjusts the volume of air provided. The air damper is vital in achieving combustion efficiency, as the motor is a constant-speed device.
Ambient temperature – the temperature of the surrounding air, e.g. boiler room air temperature.
Class C2 oil – a grade of distillate oil, also known as kerosene. It has a viscosity of 28 seconds.
Class D oil – a grade of distillate oil, very similar to the oil used in diesel engines. It has a viscosity of 32 seconds.
CO_2 indicator – a special instrument used to establish the carbon dioxide content of the products of combustion, an important factor in determining the efficiency of the boiler. It consists of a body, which has two reservoirs, one at the bottom and one at the top of the body. A hollow stem between the reservoirs has an adjustable incremental scale attached that measures the CO_2 content as a percentage. Within the body is an absorption liquid that is subjected to an increase in volume when CO_2 is absorbed. To complete the instrument there is a sampling tube that has a plunger at one end and a sampling probe at the other, and at the plunger end is an aspirator for pumping flue gases from the flue. With the probe inserted into the flue sampling hole the plunger is pushed down on to the top valve of the indicator and pumped 18 to 20 times to extract a gas sample. The sampling tube is purged prior to being used to produce a sample by pumping the aspirator six times. After the sample has been taken the indicator body is inverted several times to allow the liquid and gas to mix thoroughly. At the end of this process the level of the expanded liquid will indicate the percentage volume of CO_2 in the flue gases.
Combustion air – as with any fuel-burning appliance air has to be provided to ensure complete combustion takes place during the combustion process. In the case of oil combustion it is taken that an open-flued appliance of 5 kW or less maximum heat input does require additional ventilation. For every kW in excess of 5 kW, free-area ventilation of 5.5 cm^2 must be provided. If a draught stabilizer is installed in the flue, then the ventilation requirements must be doubled.

Example 6.1 Ventilation for a 15 kW boiler is

$$(15 - 5) = 10 \times 5.5 = 55 \text{ cm}^2 \text{ free-area vent}$$

Combustion efficiency – the efficiency of the appliance's combustion is essential to give optimum performance. Efficiency should be measured on commissioning an appliance and at every periodic service. The combustion efficiency is achieved by analysing the following readings

i the amount of flue draught measured in mbar or pascals;
ii the amount of particulates of carbon present in the products of combustion – measured against a chart;
iii the amount of carbon dioxide (CO_2) as a percentage of the products;
iv the net flue gas temperature; flue gas temperature—ambient temperature;
v the oil pump pressure.

The skilled service engineer can manipulate the burner to give the best efficiency based on the manufacturer's instructions.
Contents gauge – either a clear plastic tube on the outside of the tank or alternatively a float-operated gauge mounted on the top of the tank with a float suspended on nylon string inside the tank. Contents gauges are a useful means of determining the actual capacity of the tank.
Control box – internal electronic circuits which control a predetermined sequence of events to coordinate the burner components to bring about combustion and its safe operation. If a component malfunctions the sequence is aborted and a 'lock-out' switch is operated. A manual reset button must be pushed to restart the burner.
Dead weight and fusible link fire valve – a type of fire valve that incorporates a valve mechanism, which is lever operated. The lever is loaded with a dead weight, which is suspended by a steel cable in the open position. The cable runs through small pulleys fixed to the boiler-house walls in a position where it passes over the boiler burner. Directly above the burner is a fusible link interrupting the cable.
Distillate oils – oils used in domestic and light commercial heating boilers. They are distilled from crude oil at the refinery by heating the crude oil to very high temperatures and then condensing the resultant vapours at various densities to produce different grades of oil.

Table 6.1 Properties of distillate oils

Property	Class C2	Class D
Colour	Clear	Red
Calorific value	44.8 MJ/kg (46.3 kW/gal.)	45 MJ/kg (48.1 kW/gal.)
Relative density	0.79	0.83
Viscosity	28 s	32 s
Solidification temperature	−40°C	−10°C
Flash point	38°C	55°C
Sulphur content	0.2%	1.0

Drain point – a means of draining off the contents of the tank or alternatively the accumulations of debris or water, which may have entered the tank. It is usual for a 20 mm female BSPT thread to be provided which should have a plugged gate valve connected to it.
Draught stabiliser – effectively a centrally hinged flap valve, which has a counterweight or balance attached to the flap. It is situated at the base of the flue usually as close to the appliance as is practicable. Its function is to maintain the desired draught value. When adverse conditions occur in the flue the flap opens to admit air or conversely to expel excess pressure in the flue gases.
Electric motor – normally an inductive single-phase motor provided to give the motive force to drive the air fan and the oil pump simultaneously. The control box activates the motor.
Electrodes – there are two electrodes, which provide a source of ignition by producing a high voltage arcing across a predetermined gap. When the burner is switched on the control box sequence activates the arcing prior to the fuel starting to flow from the nozzle. The electrodes are effectively two conductors carefully positioned adjacent to the nozzle.
Electronic flue gas analyser – the modern approach to commissioning and setting up oil burners is to use an electronic flue gas analyser. This type of machine enables the engineer to adjust the burner components whilst the analyser is operating, therefore making the task much simpler. The machine has a probe at the end of a suction hose that includes a thermometer and a sampling tube. Most analysers will give the percentage values for CO_2 and oxygen content, net flue gas temperature and the efficiency of the burner. It will not give the soot content and therefore this value still requires the use of a smoke pump.
Filling inlet – a means by which the oil supplier replenishes the oil level in a tank. The inlet should be 50 mm diameter and have a 50 mm BSPT male thread available for connection to the supply tanker's hose. The inlet should have a purpose-made brass cap screwed on to it except when filling is taking place. A short chain attaches the cap to the inlet.
Fire valve – an extremely important part of the supply pipework system. Its function is to cut off the fuel supply in the event of a fire at the boiler site. There are three basic types; dead weight and fusible link, fusible wheelhead and the pressure type.
Flue draught gauge – an instrument that is used to measure the magnitude of the flue draught. It will have a calibrated scale in increments of inches (water gauge) or latterly in mbar. The scale measures in negative and positive values. Before it is used, the gauge is 'zeroed' by means of a lever protruding from its back. By inserting the attached probe into the flue pipe and observing the gauge

the user can read off the value. The value of the flue draught is then compared with the manufacturer's data, which specifies the desired draught pressure. This is used to establish the performance of the flue.
Flue gas thermometer – an instrument specially made to withstand high temperatures found in flueways. It is capable of measuring high temperatures, when inserted into the sampling hole in the flue pipe. The flue gas temperature is used in calculating the overall efficiency of the boiler. The thermometer measures the gross flue gas temperature. From this the ambient temperature of the boiler room is subtracted to arrive at the net flue gas temperature.
Flue liner – a liner constructed from acid-resistant stainless steel, spirally formed to give it flexibility for ease of insertion into Class 1 chimneys (flues designed for solid fuel combustion). The chimney must be sealed both at the top and the bottom with a suitable flue terminal placed at the outlet of the flue. It is recommended that the space between the flue liner and the chimney be filled with 'loose fill insulation'. This prevents the condensation of the products of combustion and the production of corrosive acids.
Fuel consumption – the amount of oil consumed by an appliance. It can assist in predicting the capacity of an oil tank. The calculation to determine the fuel consumption of a boiler is dependent on the following information: boiler output and efficiency, the calorific value of the oil, and the relative density of the oil. The following equation may be used

$$\text{Oil consumption} = \frac{\text{Boiler output} \times 3.6}{\text{Eff\%} \times \text{CV} \times \text{RD}}$$

where:

3.6	=	megajoules per kilowatt	(MJ/kW)
Eff%	=	efficiency as a percentage	(%)
CV	=	calorific value of the oil	(MJ/kg)
RD	=	relative density of oil	(kg/m^3)

Example 6.2 An oil-fired boiler is required to have an output of 20 kW. The boiler has an efficiency of 80% and is set up to burn Class C2 oil (CV 44.8 MJ/kg) which has a relative density of 0.79. Calculate the oil consumption per hour of operation.

$$\text{Oil consumption} = \frac{\text{Boiler output} \times 3.6}{\text{Eff\%} \times \text{CV} \times \text{RD}}$$

$$= \frac{20 \times 3.6}{0.8 \times 44.8 \times 0.79}$$

$$= \frac{72}{28.3}$$

$$= \underline{\textbf{2.54 litres/h}}$$

If the boiler is expected to run 10 hours per day at design conditions the daily consumption will be

$10 \times 2.54 =$ **25.4 litres/day**

Therefore a 2700 litre tank would provide a minimum uninterrupted supply of

$\frac{2700}{25.4} =$ **106 days or 3.5 months**

Example 6.3 An oil boiler has an output rated at 15 kW when operating with an efficiency of 82%. Calculate the daily consumption based on a 12 hour daily heating period when burning Class D heating oil.

$$\text{Oil consumption} = \frac{\text{Boiler output} \times 3.6}{\text{Eff\%} \times \text{CV} \times \text{RD}}$$

$$= \frac{15 \times 3.6}{0.82 \times 45 \times 0.83}$$

$$= \frac{54}{30.63}$$

$$= \mathbf{1.76\ litres/h}$$

Therefore daily consumption is

$12 \times 1.76 \quad = \mathbf{21.12\ litres}$

Fuel pump – a pump for drawing the oil from the storage tank. Alternatively the oil may be delivered to it at atmospheric pressure by gravity flow. The pump increases the pressure of the oil to 7 or 8 bar depending on the fuel type. These pressures are necessary to force the oil through the minute orifice of the nozzle. The fuel pump always pumps a constant amount of oil regardless of the amount of oil being burnt at the burner nozzle.

Fusible link – a device for incorporating into the steel cable which supports the weight of a dead-weight fire valve. It consists of two brass plates soldered together with special low-melting-point solder. On one end of the two plates an eyelet is formed to attach on to the steel cable. In the event of a fire occurring the solder quickly melts separating the two plates which then causes the dead weight to drop cutting off the oil supply.

Fusible wheelhead fire valve – essentially a wheelhead stop valve, which under normal circumstances can function as an isolation valve. The valve mechanism is spring-loaded so that when in the open position it is tending to slam shut. The wheelhead and the valve spindle are threaded. The internal thread of the wheel is cut through a low melting point solder. In the event of a fire the solder melts and releases the valve spindle from the wheelhead enabling the force of the spring to shut off the oil supply.

High-tension leads – special cables which connect the electrodes to the transformer. They are similar to the leads used on petrol-driven car engines.

Nozzle – a finely engineered component with an integral filter, which atomizes the oil in preparation for ignition. The nozzle will spray the oil out into the combustion zone in a predetermined angle dependent upon the length and width of the combustion chamber, typical angles of spray being 45°, 60° and 80° in domestic appliances.

Oil filter – a pipeline device which is inserted into the oil supply pipe either immediately downstream of the oil stop valve attached to the storage tank or adjacent to the appliance. It consists of an inlet and outlet connection and a bowl containing a disposable paper cartridge filter element, which the oil must flow through. Its function is to remove scale and deposits from the oil tank before the oil reaches the burner.

Oil pipeline connection – a connection located near the base of the tank on the opposite side to the drain. Normally a 15 mm female BSPT thread is provided. It is good practice to provide an isolating valve close to the connection for maintenance or emergency work.

Oil pipeline materials – there are two common pipe materials used to install oil pipework. Firstly, low-carbon steel pipe manufactured to BS 1387 jointed using pipe threads to BS 21. The preferred jointing medium is a non-setting paste such as 'red haematite'. Secondly, copper tube manufactured to BS 2871: Table Y – 'soft copper'. The preferred jointing method is the use of manipulative compression fittings. Capillary joints should be avoided. For most domestic installations 10 mm diameter copper tube is adequate.

Oil pressure gauge – a pressure gauge of the Bourdon gauge type, which is attached to an extension piece containing an isolation valve. A pressure sampling port is built in to the oil pump of a pressure jet burner for which the gauge assembly is inserted. It allows the engineer to record the initial pressure and also adjust the actual oil pressure to suit the type of oil being burnt. The information regarding oil pressure is detailed in the manufacturer's data sheet.

Oil storage tank – a tank to store oil on site in readiness for use. Storage tanks were traditionally manufactured from low-carbon steel sheets arc welded together, either in a workshop situation or on site. In recent times we have experienced the introduction of corrosion free plastic tanks. Storage tanks are available in various capacities.

One-pipe oil supply – a one-pipe oil supply system (also known as a gravity system) may be used to supply oil to both a vaporizing

and an atomizing oil burner when the vertical height between the base of the tank and the burner is no less than 0.3 metres and no greater than 3 metres. Essential pipeline components should include isolation valves, an oil filter and a fire valve.
Open vent – a vent to maintain atmospheric pressure within the tank during filling and emptying (when oil is being drawn off). The vent should be 50 mm diameter with a means of preventing the ingress of rainwater and vermin. Purpose-made vents are available from suppliers.
Photoelectric cell – a flame failure device in a burner. It consists of a light-sensitive resistor, which when receiving light enables a circuit to send a millivolt signal to the control box. It is inserted into the air supply tube facing the combustion chamber. If the flame is not established or becomes extinguished it causes the control box to go to 'lock-out'.
Pressure jet burner – a type of burner which consists of several components that collectively support combustion by atomizing the oil into minute droplets. The burner is made up of the following component parts

air diffuser
air fan
control box
electrodes
electric motor
fuel pump
high-tension leads
nozzle
photoelectric cell
solenoid valve
transformer.

Pressure type fire valve – a fire valve whose operation relies on a heat-sensitive vapour contained within a phial and an expanding bellows; the two elements are connected by a capillary tube. When the vapour is subjected to intense heat, from a fire for instance, the vapour expands rapidly. The expanded volume of vapour exerts pressure in the bellows, which in turn closes the fire valve. The phial must be in the same room as the appliance, preferably above the burner.
Smoke pump – an instrument for sampling the amount of free carbon particulates in the products of combustion. It is a similar device to a bicycle pump with a few modifications; attached to the suction end is a stainless steel probe, which is inserted into the flue pipe. A special filter paper is placed in a slot in the end of the pump body and secured by turning down a knurled disc. When the probe has been inserted into the flue the pump is pumped ten times; this enables an adequate sample to be taken. The resultant circle of discoloration is then held against a 'smoke scale' to establish the extent of free carbon.
Solenoid valve – essentially an electromagnetically operated valve, which is opened automatically by the control box during the ignition cycle. It is opened after the ignition arcing has commenced. It is situated inline between the fuel pump and the nozzle. It also enables the pump to build up the oil operating pressure whilst the solenoid is closed.
Storage capacity – the minimum practical capacity is deemed to be 1250 litres, however for economic reasons in terms of suppliers' delivery charges it is desirable to have a storage capacity of 2700 litres.
Tank installation – storage tanks may be situated externally or internally to a building, the latter being much more expensive due to safety factors. Tanks should have the following features: a sound support, an open vent, a filling inlet, a contents gauge, an oil pipeline connection, a drain point and a gradient of 20 mm/m of length, falling towards the drain point.
Tank supports – these should be brick or concrete blockwork piers built on a concrete foundation, which provide a suitable gradient for draining purposes and also the prevention of particles or grit entering the supply pipework.
Tiger-loop deaerator – a diaphragm valve which has attached to it a float chamber containing an automatic air release valve, maintaining a constant oil level within the chamber. Therefore the oil pump, which can have a pumping capacity up to 20 times the actual capacity of the oil being burnt, can circulate the excess oil between the Tiger loop and the pump several times before it is burnt, enabling efficient deaeration of the oil. Therefore in effect the only oil that is sucked out from the tank is that which is being burnt. The Tiger loop can enhance performance and efficiency.
Transformer – a device for changing voltage. To produce a temperature in the arc across the two electrodes sufficient to ignite the oil it is necessary to increase the supply voltage from 240 V to 10 000 V. This function is performed by the transformer.
Two-pipe oil supply – when the base of the oil storage tank is less than 0.3 m above the burner a two-pipe system should be installed. This requires the installation of a suction pipe from the base of the tank to the oil pump and a return pipe from the 'return

port' on the oil pump to a dip-pipe which enters the top of the tank and terminates close to the tank base – this is to keep the system free of air. Additional pipeline components include a non-return valve to prevent oil flowing back to the tank.

Vaporizing oil burners – a type of appliance in which the oil for combustion is pre-heated to a temperature where it can change into its gaseous state. This enables it to mix with oxygen to produce a combustible mixture. Ignition can be self-perpetuating or by means of an arc across two electrodes by a voltage of 10 000 V. This type of burner is less favoured for heating boilers.

Viscosity – the property of a liquid to flow. The higher the density of a liquid the greater will be the value of its viscosity. Viscosity is measured using a device known as the Redwood N° 1 viscometer. A measure of the liquid (50 cm^3) at a constant temperature of 37°C is timed, in seconds, to establish how long it takes to pass through a defined orifice diameter. Heavy fuel oil for instance takes 3600 seconds whereas Class D oil takes only 32 seconds.

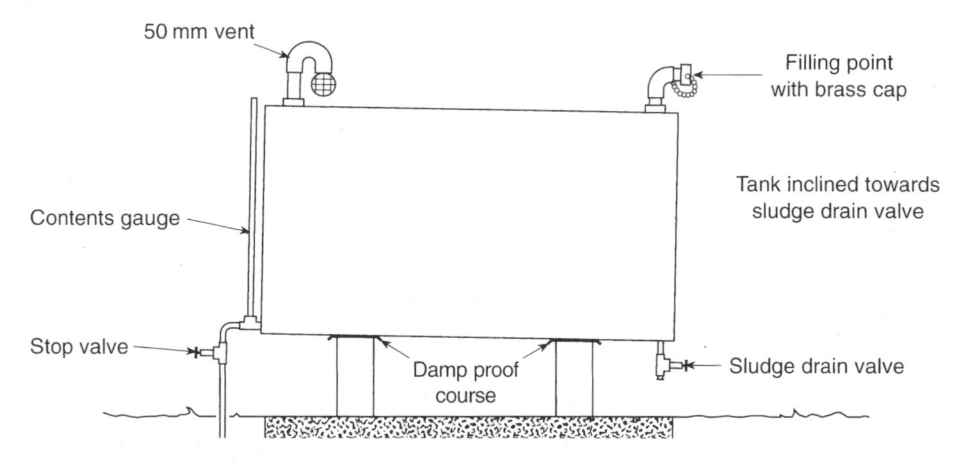

Fig. 6.1 External oil tank installation

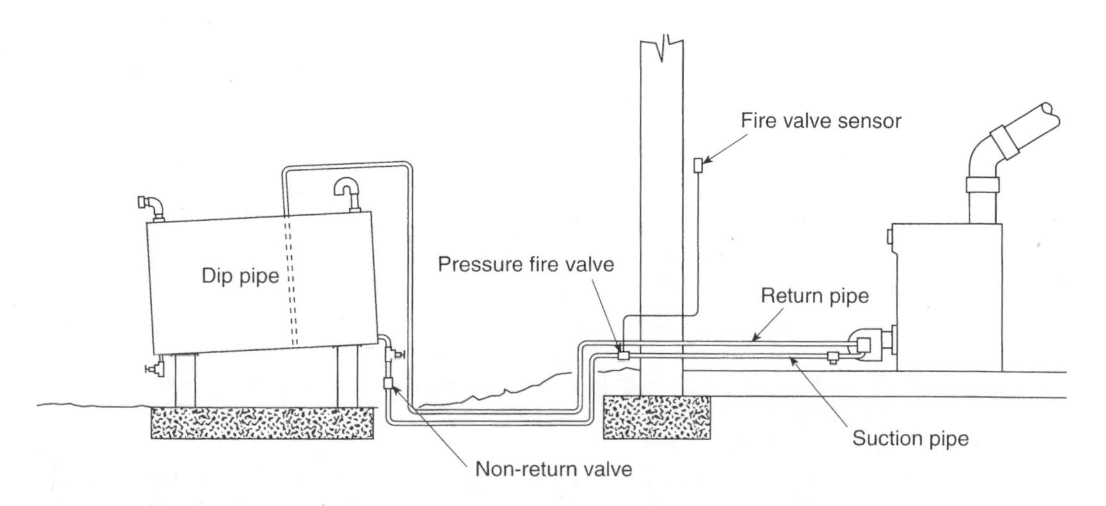

Fig. 6.2 Two-pipe oil supply

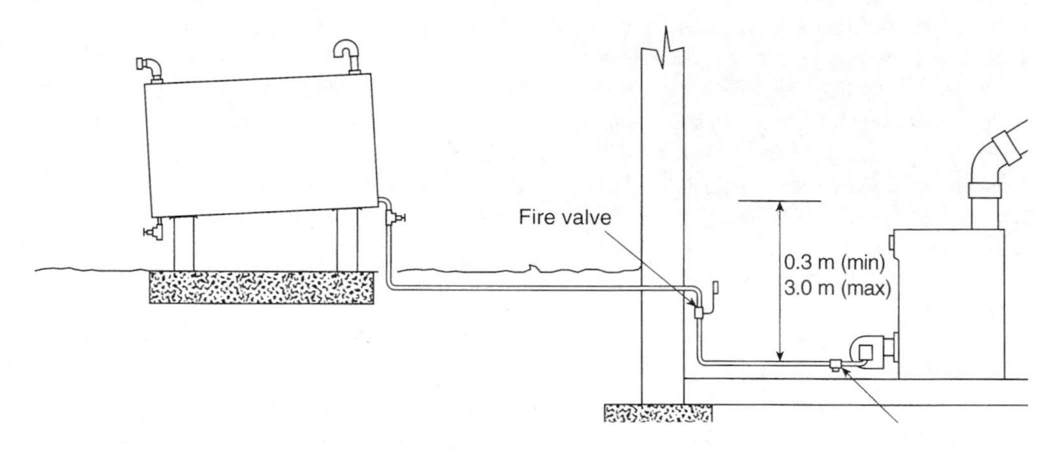

Fig. 6.3 One-pipe oil supply

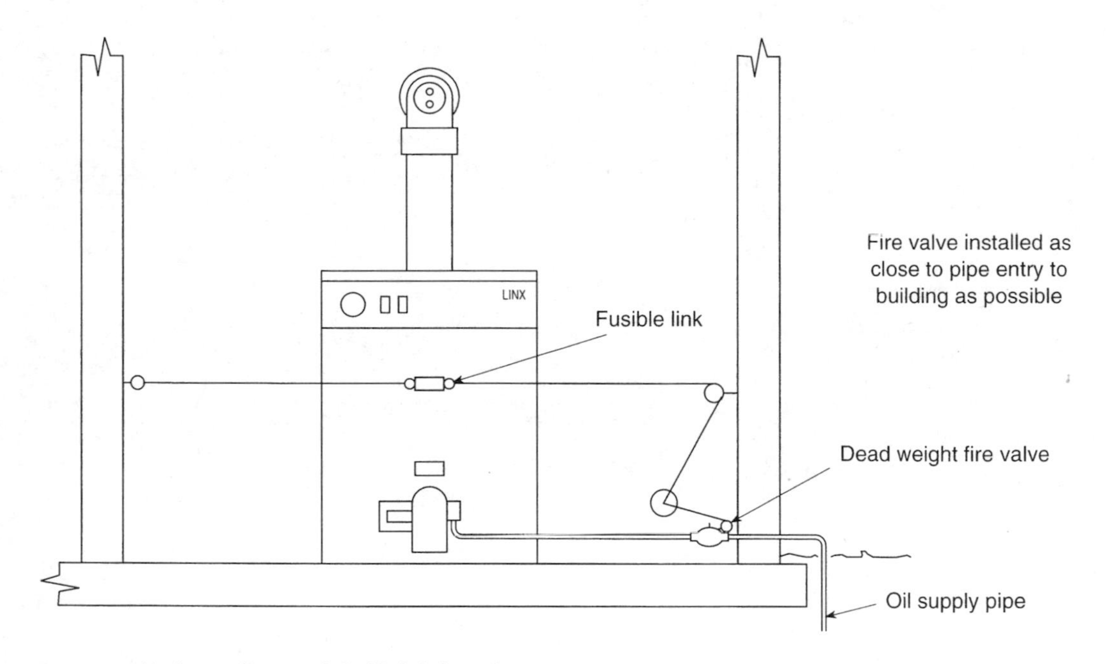

Fig. 6.4 Oil boiler installation with fusible link fire valve

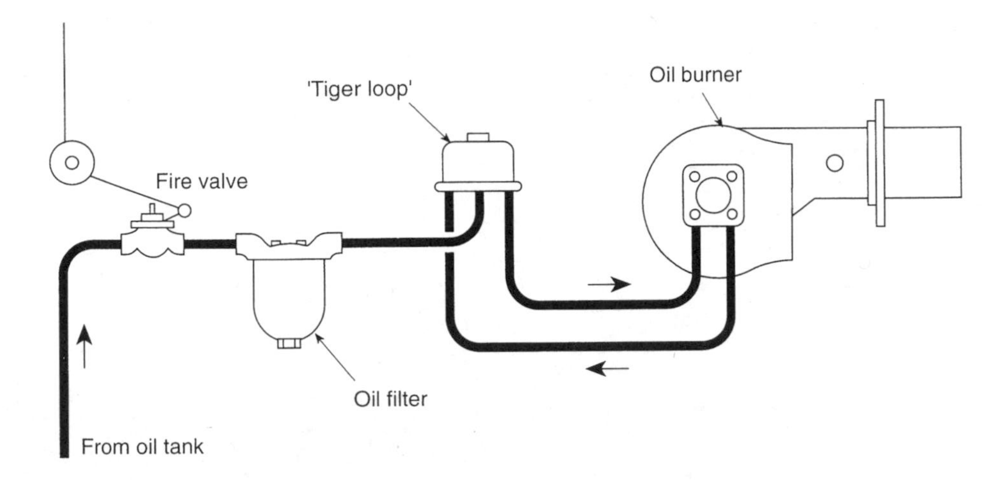

Fig. 6.5 'Tiger loop' deaerator installation

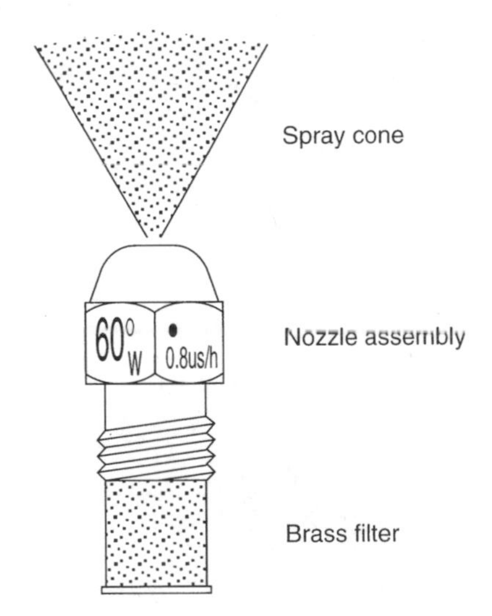

Fig. 6.6 Pressure jet burner nozzle

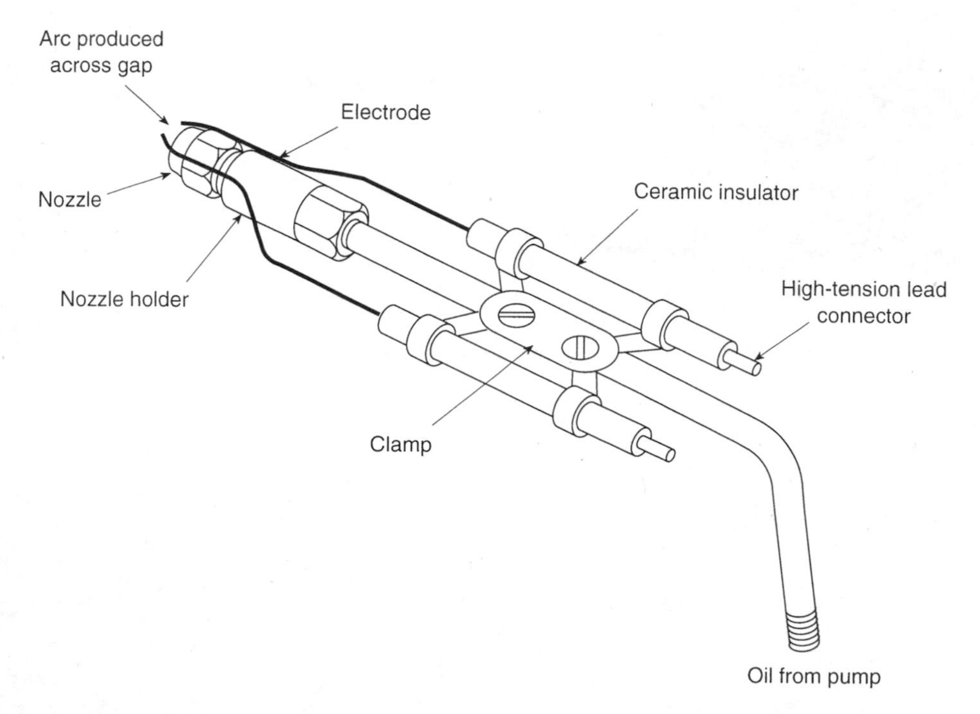

Fig. 6.7 Nozzle and electrode assembly

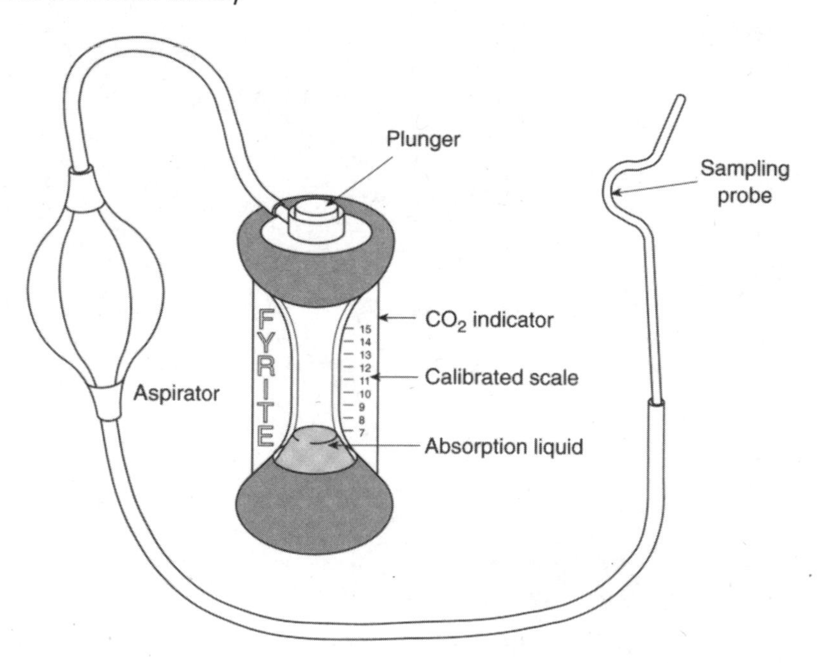

Fig. 6.8 CO_2 indicator

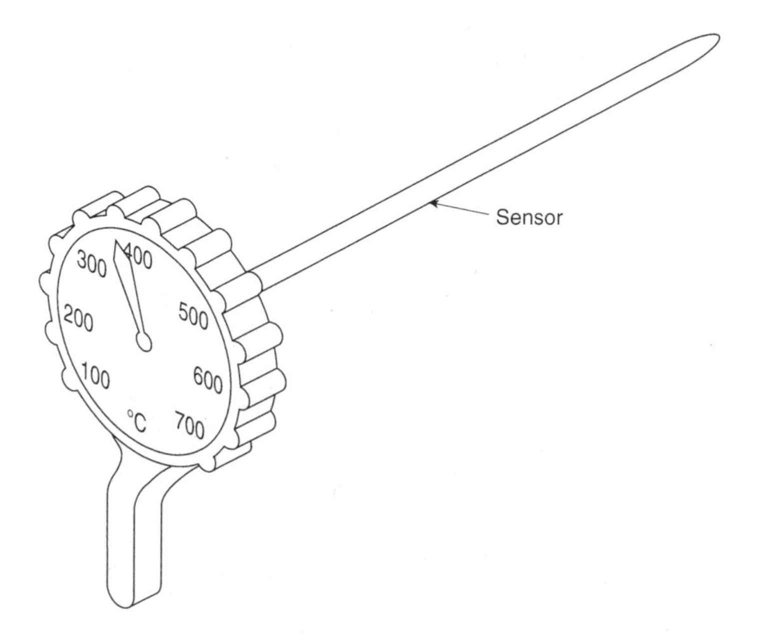

Fig. 6.9 Flue gas thermometer

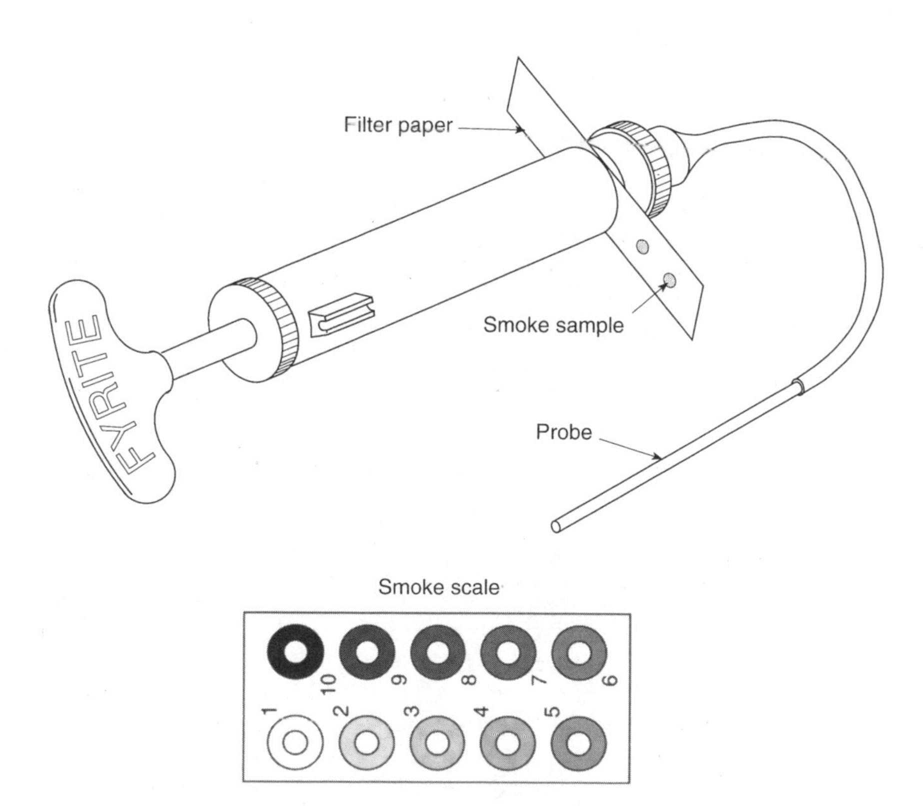

Fig. 6.10 Smoke pump

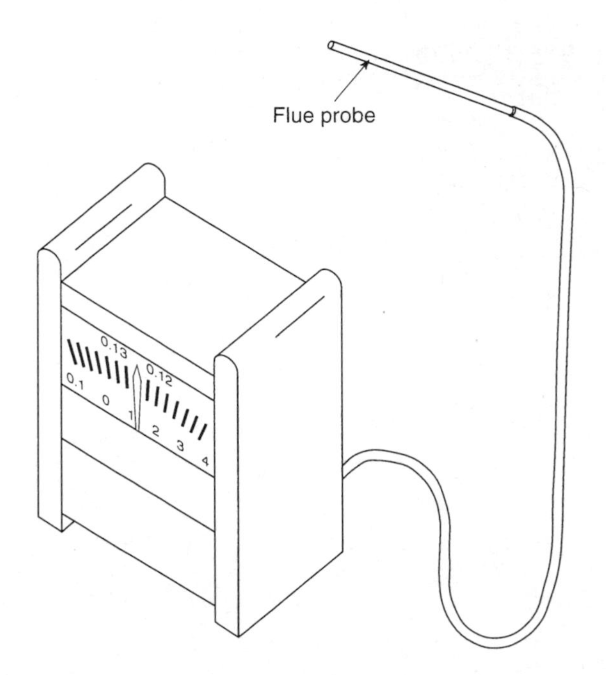

Fig. 6.11 Flue draught gauge

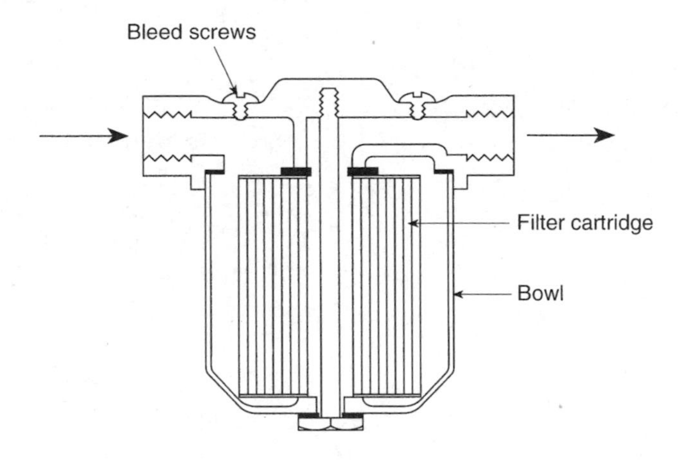

Fig. 6.12 Oil filter

Now try and answer these questions

1. The classification of oil used in domestic heating appliances is referred to as

a heavy fuel oils
b lubricating oils
c crude oils
d distillate oils.

2. The relative density of Class C2 oil is

a 1.00
b 0.79
c 1.05
d 0.83.

3. Class D domestic fuel oil has a viscosity of

a 32 seconds
b 28 seconds
c 60 seconds
d 90 seconds.

4. Viscosity is the property of a liquid to

a boil
b adhere
c freeze
d flow.

5. An oil filter is a pipeline component whose purpose is to prevent scale, etc. from reaching the

a oil tank
b fire valve
c burner
d flue pipe.

6. The part of an oil filter which prevents scale passing into the system is a

a paper cartridge element
b metallic gorse mesh
c wedge of paper towels
d porous stone plug.

7. A fire valve is a device on an oil pipeline which will automatically __________ in the event of a fire.

a extinguish the burner
b reduce the flow of oil
c return the oil to the tank
d cut off the oil supply.

8. A type of fire valve which relies on a link held together with a low-melting-point solder is referred to as a

a fusible wheelhead fire valve
b dead weight and fusible link fire valve
c pressure type fire valve
d fire stop valve.

9. A type of oil burner which generates high pressure in the oil and forces it through a fine orifice to atomize it is called

a a vaporising burner
b an evaporation burner
c a pressure jet burner
d a pot burner.

10. A device known as an 'air diffuser' is an important factor in maintaining the __________ in pressure jet burners.

a flame stability
b induction of air
c filtering of combustion air
d reduction of burner noise.

11. The type of fan impeller used on a pressure jet burner is a

a propeller impeller
b axial impeller
c bifurcated impeller
d centrifugal impeller.

12. On the air inlet of a pressure jet burner fan is a damper. This is used to maintain

a excess air supply
b stoichiometric combustion
c fan speed
d boiler combustion efficiency.

13. The control box of a pressure jet burner ______________ maintains burner combustion when the system calls for heat.

a mechanically
b automatically
c instinctively
d positively.

14. When a component of a pressure jet burner malfunctions the control box goes to 'lock-out'. This can only be corrected

a manually
b automatically
c by replacement
d by a specialist.

15. The flame failure device of a pressure jet burner is referred to as a

a bimetallic rod
b thermocouple
c photoelectric cell
d vapour expansion device.

16. The transformer on a burner increases the supply voltage of 240 V to ________ to produce the arc across the electrodes.

a 1100 volts
b 10 000 volts
c 5000 volts
d 20 000 volts.

17. Copper tube used in the installation of oil pipelines should be

a BS 2871 Table X
b BS 2871 Table W
c BS 2871 Table Z
d BS 2871 Table Y.

18. For domestic oil installations the most appropriate diameter is

a 8 mm
b 15 mm
c 10 mm
d 22 mm.

19. Low-carbon steel tube used for oil supplies should be manufactured to

a BS 596 Table 1
b BS 1387
c BS 1192
d BS 59: Part 3.

20. Threads on low-carbon steel pipe should be cut to

a BS 21
b BS 450
c BS 1553
d BS 6644.

21. Oil is kept in readiness for use in a vessel referred to as a

a oil storage tank
b oil reservoir
c bulk receiver
d oil drum.

22. An oil tank installation should incorporate a drain point with a gradient falling towards the drain of

a 1 in 600
b 1 in 3
c 20 mm/metre of length
d 50 mm/metre of length.

23. A tank contents gauge enables the user to determine the

a type of oil in the tank
b actual contents of the tank
c theoretical contents of the tank
d amount of water in the tank.

24. The purpose of an open vent on top of an oil tank is to maintain

a atmospheric pressure
b the oil level
c combustion
d a corrosion free zone.

25. One advantage of using a fusible wheel-head fire valve is that it can also function as a

a mixing valve
b diverting valve
c float-operated valve
d isolating valve.

26. The application of the pressure type fire valve enables the valve to be located outside the room containing the

a oil tank
b capillary tube
c oil filter
d sensing phial.

27. The electrodes in a pressure jet burner are the source of ignition. They are activated

a continuously during combustion
b at the same moment oil flows
c before the oil starts to flow
d after the oil has started to flow.

28. The nozzle of a pressure jet burner is the component which ________ the oil.

a atomizes
b vaporizes
c combusts
d pressurizes.

29. A solenoid valve is an isolating device which is operate

a manually
b electromagnetically
c intermittently
d constantly.

30. The minimum practicable oil storage capacity is deemed to be

a 1250 litres
b 3000 litres
c 2700 litres
d 1000 litres.

31. When installing oil fired heating, it is desirable to have oil storage of at least

a 1250 litres
b 3000 litres
c 2700 litres
d 1000 litres.

32. The oil delivery tanker has a filling hose with a ________ female thread connector.

a 100 mm
b 40 mm
c 25 mm
d 50 mm.

33. One of the factors required in assessing the daily oil consumption of an appliance is the

a output
b manufacturer
c flue dimensions
d oil tank location.

34. An oil boiler has an output of 12 kW with an efficiency of 83%. If it is burning Class C2 oil its hourly consumption will be

a 1.47 litres
b 2.47 litres
c 1.57 litres
d 2.57 litres.

35. If a boiler has an output of 35 kW at an efficiency of 80% whilst set up to burn Class D oil, the hourly consumption will be

a 1.22 litres
b 4.22 litres
c 4.88 litres
d 1.88 litres.

36. A device when fitted in the oil supply pipeline adjacent to the burner that will enhance performance and efficiency is referred to as a

a fuel injector
b pressure governor
c tiger-loop deaerator
d constant pressure governor.

37. An open-flued oil appliance has a maximum heat input of 35 kW. The size of free air ventilation required will be

a 150 cm^2
b 265 cm^2
c 250 cm^2
d 165 cm^2.

38. An open-flued oil fired boiler with a maximum heat input of 20 kW with a draught stabilizer installed in the flue would require free air ventilation of

a 150 cm^2
b 265 cm^2
c 250 cm^2
d 165 cm^2.

39. Flue liners for oil appliances are manufactured from

a stainless steel
b low-carbon steel
c heat-resistant plastics
d salt glazed earthenware.

40. Combustion efficiency is achieved by analysing the

a chemical properties of oil
b products of combustion
c soot content of the combustion zone
d the whole boiler installation.

41. An instrument known as a smoke pump is used to measure the ________ in the products of combustion.

a carbon dioxide percentage
b vapour pressure
c carbon particulates
d spillage.

42. When assessing burner efficiency the ________ temperature is deducted from the flue gas temperature.

a ambient
b return pipe
c flow pipe
d heating design.

43. The fuel pump on a pressure jet burner is capable of increasing the pressure of the oil up to

a 2 or 3 bar
b 7 or 8 psi
c 2 or 3 psi
d 7 or 8 bar.

44. The instrument used to assess the carbon dioxide content of products of combustion is referred to as a

a SO_4 gauge
b H_2O indicator
c CO_2 indicator
d CH_4 gauge.

45. A disadvantage of using only an electronic flue gas analyser when setting up oil burners is that it cannot

a measure the soot content
b detect the oxygen content
c calculate the net stack temperature
d give the engineer a print out.

46. An oil-fired boiler has a rated output of 22 kW at an efficiency of 85%. If the burner is set up to burn Class C2 oil and be expected to run for 10 hours per day, calculate the capacity of an oil storage tank required to provide an uninterrupted supply for six weeks.

47. Calculate the number of days uninterrupted supply of oil that could be achieved from a storage tank with an actual capacity of 2000 litres for the following situation.

Data:

Boiler	–	pressure jet burner, 18 kW heat output
System efficiency	–	82%
Oil type	–	Class D
Daily operation	–	8 hours

Now check your answers from the grid.

Q 1; d	Q 10; a	Q 19; b	Q 28; a	Q 37; d
Q 2; b	Q 11; d	Q 20; a	Q 29; b	Q 38; d
Q 3; a	Q 12; d	Q 21; a	Q 30; a	Q 39; a
Q 4; d	Q 13; b	Q 22; c	Q 31; c	Q 40; b
Q 5; c	Q 14; a	Q 23; b	Q 32; d	Q 41; c
Q 6; a	Q 15; c	Q 24; a	Q 33; a	Q 42; a
Q 7; d	Q 16; b	Q 25; d	Q 34; a	Q 43; d
Q 8; b	Q 17; d	Q 26; d	Q 35; b	Q 44; c
Q 9; c	Q 18; c	Q 27; c	Q 36; c	Q 45; a

SOLUTION TO QUESTION 34

$$\text{Oil consumption} = \frac{\text{Boiler output} \times 3.6}{\text{Eff\%} \times \text{CV} \times \text{RD}}$$

$$= \frac{12 \times 3.6}{0.83 \times 44.8 \times 0.79}$$

$$= \frac{43.2}{29.37}$$

$$= \mathbf{1.47\ litres/hour}$$

SOLUTION TO QUESTION 35

$$\text{Oil consumption} = \frac{\text{Boiler output} \times 3.6}{\text{Eff\%} \times \text{CV} \times \text{RD}}$$

$$= \frac{35 \times 3.6}{0.8 \times 45 \times 0.83}$$

$$= \frac{126}{29.88}$$

$$= \mathbf{4.22\ litres/hour}$$

SOLUTION TO QUESTION 46

Oil Consumption $= \dfrac{\text{Boiler output} \times 3.6}{\text{Eff\%} \times \text{CV} \times \text{RD}}$

$= \dfrac{22 \times 3.6}{0.85 \times 44.8 \times 0.79}$

$= \dfrac{79.2}{30.08}$

= **2.63 litres/hour**

Daily consumption = l/hr × hr/day
= 2.63 × 10
= **26.3 litres**

Storage capacity = Daily consumption × number of days
= 26.3 × (6 × 7)
= **1104.6 litres**

SOLUTION TO QUESTION 47

Oil consumption $= \dfrac{\text{Boiler output} \times 3.6}{\text{Eff\%} \times \text{CV} \times \text{RD}}$

$= \dfrac{18 \times 3.6}{0.82 \times 45 \times 0.83}$

$\dfrac{64.8}{30.63}$

= **2.12 litres/hour**

Daily consumption = l/hr × hr/day
= 2.12 × 8
= **16.96 litres**

Number of days supply $= \dfrac{\text{tank capacity}}{\text{litres/day}}$

$= \dfrac{2000}{16.96}$

= **117.78 days** **say 118 days**

7

ELECTRICAL SUPPLY

To tackle this chapter you will need to know:

- types of electricity generated;
- distribution of electricity;
- cables for conducting electricity;
- electrical wiring circuits;
- circuit protective devices;
- electrical installation.

GLOSSARY OF TERMS

Alternating current (AC) – the type of electricity supplied through the mains cables entering a building. Basically the voltage and current alternate between peaks of positive and negative values. These coincide with the rotation of the generator. The current lags behind the voltage.

Ampere (A) – the unit of flow of negative charges (electrons) around a circuit or along a conductor. An analogy is the case of a cistern full of water connected via a pipe to a valve. With the valve closed there is no water flowing, but when the valve is opened the water flows. The velocity of flow is dependent on the height (pressure) of the tank and the cross-sectional area of the pipe. Therefore current is dependent on voltage (pressure) and cable size.

Circuit protective devices – effectively these are safety devices which sense current overload and react by isolating the supply automatically. They may take the form of rewirable fuses (BS 3036), cartridge fuses (BS 1361; BS 1362; BS 88) or circuit breakers.

Conduit – steel or plastic tubing systems for electrical wiring installation jointed by screwed fittings or solvent welded fittings respectively, connecting the consumer unit to the socket outlets or light fittings. The tubing is threaded with single cables to connect the circuits. This system offers some mechanical protection to the conductors particularly in the case of steel conduits.

Consumer unit – a combined controlling and protecting facility for all the final circuits in the building. It contains a double-pole switch to isolate the all the circuits from the mains supply, a phase busbar with circuit protective devices linking the supply to the final circuits, a neutral connector block and an earth connector block enabling all final circuits to be terminated in one box.

Continuity test – an electrical test to establish that any short circuit occurring has a path to earth via the CPC and not through anyone touching the casing of an appliance or component. A direct path to earth will operate the circuit protective device. The test is carried out using a low-resistance (Ω) multimeter, one probe is attached to the earth conductor at the fused spur and the other one at various points on the appliance, e.g. earth connection of the appliance, casing, pipework, heat exchanger, etc. The reading should be less than 1 Ω.

Continuous protective conductor (CPC) – a continuous conductor which is connected to all parts of the electrical installation, including appliances, switches, conduits and the consumer unit earth block. There should be no means of isolating any section of it. The conductor is extended from the consumer unit to the main earth return conductor. In modern electrical installations the CPC is connected to the neutral link at the main cable.

Direct current (DC) – a source of electricity which maintains a constant voltage and current. A typical supply is that obtained from batteries.

Distribution – electricity is distributed throughout the UK by means of overhead cables mounted on pylons. This system is referred to as the National Grid. The controllers of the grid can divert power to wherever it is required from a central control centre. Because of the vast distances travelled the initial voltage is increased to 400 000 volts, with step-up transformers at the power station, to take care of voltage decay.

Double pole switch – a type of switch which when operated will either connect or disconnect both the phase and the neutral conductors. This, when used to disconnect a supply, has the effect of completely isolating the circuit from the supply. They are found in consumer units, immersion heater circuits and electric shower circuits.

Earth clamp – a mechanical device for connecting an equipotential bonding conductor to a permanent metallic structure such as gas pipes, water pipes, etc. It is attached by passing a strap through a clamping mechanism tightened by a screw. It has a stamped aluminium plate advising against its removal. Only a competent person should attach it.

Equipotential bonding – it is a requirement of BS 7671 that all exposed permanent metalwork in a building must be bonded together and consequentially connected to the earth block in the consumer unit. This prevents a potential difference existing in the metalwork should a stray voltage be present from short circuit. It is also necessary to bond any incoming metallic service pipes, i.e. gas, oil or water. In the case of natural gas an earth clamp must be attached within 600 mm of the meter outlet.

Final circuits – a term given to the circuits providing electrical power in a building e.g. ring mains, radial circuits and circuits for lighting.

Fixed cables – this term refers to all wiring which is a permanent feature of the installation, similar to the pipework of a central heating system. Conductors are manufactured from copper; these are usually solid wires up to 10 mm^2 cross-sectional area. For larger cables multi-stranded conductors are preferred for ease of installation. Each conductor is insulated with PVC and multiple-cored cables have a sheathing of PVC covering the conductors. These cables are identified by the colour of the insulation applied, e.g. the phase conductors are identified as red, the neutral conductor as black and the CPC as yellow and green stripes. Where three phase electricity is installed the phase conductors are identified as red, yellow and blue.

Flexible cables – cables that are used to connect socket outlets and fused spurs to appliances. They are manufactured by slightly twisting many fine strands of copper wire. This method of production enables them to be twisted and bent many times with minimal damage being caused. If the same treatment were applied to fixed cables they would break in a short period of time. The conductor is insulated and sheathed with layers of PVC. These cables are generally round in section, whereas fixed cables are flat. The amperage of the appliance governs the size of the conductor. These cables are identified by the following colour code: the phase conductor has brown insulation, the neutral conductor has blue insulation and the earth conductor has yellow and green stripes.

Frequency (hertz, Hz) – the number of revolutions of the generator measured per second. It is also the number of cycles in the AC supply when voltage rises from 0 volts to a positive maximum and then to a negative maximum before returning to 0 volts. In the UK electricity is generated at 50 cycles per second – 50 Hz.

Fused spur – a means of connecting a permanently wired appliance to a ring main. It is in effect a radial circuit connected to a ring main. It provides an additional protective device that is usually smaller than the ring circuit protection. For example a fused spur providing power for a central heating system would have a 3 amp fuse in the spur. The spur would also provide a double-pole switch isolator.

Generation – electricity is generated centrally in power stations by electricity generating companies. Power stations use the following as sources of primary energy: coal, oil, gas, and nuclear energy to produce high-pressure steam for the turbine driven generators. The generators produce AC electricity at 33 000 volts at a frequency of 50 cycles per second. Electricity is also produced using hydropower and wind power.

Insulation resistance test – a test to establish the integrity of the conductor insulation. A special test meter is used, which is capable of generating a 1000 volt or 500 volt DC charge and also of measuring high resistance in ohms (Ω). One probe is attached to the phase conductor and one to the CPC neutral conductors respectively. The test button is then pressed, and the resistance must measure more than 0.5 mΩ. Only a competent person should carry out this type of testing.

Junction Box – an electrical installation fitting that is used to connect a spur cable to a ring main cable when PVC sheathed cables are installed. It is manufactured from plastic and contains three terminals for connecting the phase, neutral and CPC conductors respectively. Also available are multiple terminal boxes for used on other types of circuits, for instance the wiring of central heating controls.

Loop-in circuit – a type of final circuit associated with wiring for lighting. A cable is taken from the consumer unit to the first light ceiling rose. The cable is then looped to the next ceiling rose and so on. A phase wire is taken from the ceiling rose to the light switch. From there a switch live wire is taken back to ceiling rose to feed the lamp holder. The neutral conductor is taken from the lamp holder to the main lighting cable.

Micro-circuit breaker (MCB) – a resettable mechanical device which is replacing the cartridge fuse in the consumer unit. It contains a tripping mechanism, which utilizes the bimetallic strip concept. When excessive current flows in the circuit the strip heats up and expansion takes place. The strip bends which activates the tripping mechanism. MCBs are reset by pushing a button. If the fault is still present it will immediately trip out again. These devices are rated by current, e.g. 5 amps or 20 amps, etc.

Neutral conductor – the neutral conductor completes the electrical circuit. It is the cable of the final circuit, which is connected to the neutral block in the consumer unit. It is identified by black insulation around the conductor wire.

Ohm (Ω) – the unit of resistance to the flow of electricity along a conductor. Using the analogy of water flowing through a pipe, transporting 10 litres of water through a 10 mm pipe in one second would create a large amount of resistance to flow. If the same 10 litres of water were to be passed through a 50 mm diameter pipe this would drastically reduce the resistance to flow. It therefore follows that resistance to flow of electricity is related to the size of conductor used. Increasing resistance will promote the conversion of electrical energy to other forms of energy such as heat or light energy.

Ohm's Law – a relationship between current, resistance and voltage. The German physicist Ohm discovered this relationship. The unit of electrical resistance is referred to as an ohm (Ω). Ohm's law states that the resistance (in ohms) of a conductor is equal to the potential difference (the voltage) across its ends divided by the current flowing (in amps).

1 watts = volts × amps
$$W = V \times A$$

2 volts = amps × ohms
$$V = A \times R$$

These two formulae can be combined to give either:

3 watts = amps² × ohms
$$W = A^2 \times R$$

4
$$\text{watts} = \frac{\text{volts}^2}{\text{ohms}}$$

$$W = \frac{V^2}{R}$$

Further transposition of these formulae will enable all problems to be solved.

Example 7.1 Calculate the fuse required to protect an immersion heater circuit rated at 3 kW connected to a 230V supply.

$$W = V \times A$$

Transposing gives

$$A = \frac{W}{V} = \frac{3 \times 1000}{230} = \underline{\mathbf{13A}}$$

Example 7.2 An immersion heater has a current of 15 amps when the voltage is 240V. Calculate the heat input and the resistance of the heating element.

$$W = V \times A$$
$$= 240 \times 15$$
$$= \underline{\mathbf{3600\ watts\ (3.6\ kW)}}$$

$$R = \frac{W}{A^2} = \frac{3600}{225} = \underline{\mathbf{16\,\Omega}}$$

Example 7.3 If 230 volts are applied to a heating element with a resistance of 1000 Ω, what will be the heat given off?

$$W = \frac{V^2}{R} = \frac{52\,900}{1000} = \underline{\mathbf{52.9\,\Omega}}$$

Parallel circuit – a circuit in which the corresponding ends of the resistors are connected together, which results in each resistor receiving the same value of voltage or potential difference (PD) but the current will be shared amongst the resistors dependent on their resistance to flow. This circuit will remain operational when a resistor fails. A typical application is the circuit used in lighting. The total resistance in a parallel circuit is the reciprocal of the sum of the reciprocals of the resistors connected:

$$\frac{1}{R_t} = \frac{1}{R_1} + \frac{1}{R_2} + \frac{1}{R_3} + \ldots \text{etc.}$$

Example 7.4 If three resistors connected in parallel have the following resistances; 10 Ω, 4 Ω and 16 Ω, what is the total resistance of the circuit?

$$\frac{1}{R_t} = \frac{1}{R_1} + \frac{1}{R_2} + \frac{1}{R_3}$$

$$= \frac{1}{10} + \frac{1}{4} + \frac{1}{16}$$

$$= 0.1 + 0.25 + 0.0625$$

$$\frac{1}{R_t} = \frac{0.4125}{1}$$

Therefore

$$R_t = \frac{1}{0.4125} = \underline{2.424\,\Omega}$$

Phase conductor – the phase (live) conductor is the name given to the cable in final circuits connected to the circuit protective device in the consumer unit. It is identified by red insulation around the conductor wire.

Polarity test – an electrical test to ensure that the phase conductor is connected to the correct terminal within an appliance or indeed a final circuit. A low-resistance (Ω) multimeter should be employed to achieve this test. One probe is attached to the fused spur's phase terminal and one to the phase terminal of the appliance connector block. The resistance measured should be less than 1 Ω. This verifies that the current cannot come in contact with an appliance casing, for example. The *Electricity at Work Regulations* state that noone can work on live electricity in relation to work, therefore this test should be carried out with the supply disconnected.

Radial circuit – a final circuit which connects a single appliance to the consumer unit, having one phase, neutral and CPC conductor connected between both. A typical example is the circuit feeding an immersion heater in a hot water storage cylinder. A double-pole switch isolates this example.

Rectifier – an electrical component used to convert an alternating current to a direct current. It consists of two metallic plates (diode). The diode has a high resistance to current flow in one direction and a low resistance in the other direction. It will therefore only conduct electricity in the positive half of the cycle in an alternating current cycle.

Regulations – these are either Statutory (legally binding) or Non-statutory (codes of practice). Some examples of Statutory Regulations are

i *Health and Safety at Work Act 1974* provides the legislative framework to promote, stimulate and encourage high standards of health and safety at work

ii *Electricity at Work Regulations 1989* requires precautions to be taken against the risk of death or personal injury from electricity in work activities

iii *Reporting of Injuries, Diseases and Dangerous Occurrences Regulations (RIDDOR) 1985* requires injuries, diseases and occurrences in specified categories to be notified to the relevant authority.

Some examples of Non-statutory Regulations and other safety publications are

i *IEE Regulations 16th Edition 1992* (BS 7671) designed to protect persons, property and livestock against hazards arising from electrical installations
ii *HSE Guidance Note GS 27 Protection against Electric Shock*
iii *IEE Guidance Note 2 Isolation and Switching (2nd Edition).*

Residual current device (RCD) – a circuit protective device, which constantly monitors the current flow in a circuit in both the phase and neutral conductors within an alternating current system. When the system is fault-free the device is in balance, but if a fault to earth occurs an imbalance in the current flow causes a current to be induced in a supplementary circuit which operates an electromagnetic switch to isolate the circuit from the supply.
Ring main circuit – a type of final circuit that provides power to socket outlets used for portable appliances. It consists of a loop of cable of which both ends are connected as follows: both of the phase conductor's ends are connected to the circuit protective device, both of the neutral conductor's ends are connected to the neutral block and both of the CPC's ends are connected to the earth block within the consumer unit. A 30 amp protective device protects this circuit.
Series circuit – a circuit in which the current passes through any resistors connected consecutively. This has the effect of diminishing the voltage or potential difference (PD) as it passes through each resistor. A disadvantage of this circuit is that when a resistor fails the circuit is broken, resulting in loss of power. The total resistance in a series circuit is the sum of the resistors connected, therefore:

$$R_t = R_1 + R_2 + R_3 + \ldots \text{etc.}$$

Example 7.5 If three resistors connected in series have the following resistances; 10 Ω, 4 Ω and 16 Ω, what is the total resistance of the circuit?

$$\begin{aligned} Rt &= R_1 + R_2 + R_3 \\ &= 10 + 4 + 16 \\ &= \underline{\mathbf{30\ \Omega}} \end{aligned}$$

Service cut-out device – a device used when the incoming mains supply is connected to the internal wiring. The phase conductor is connected via a cartridge fuse. The neutral conductors are connected through a solid metal link. The function of this device is to protect the mains from excessive overload.
Short-circuit test – an electrical test to ensure that the circuits of the system or appliance have no current flowing when the switches are open (no contact). A low-resistance (Ω) multimeter is connected to the phase and neutral conductors. When all switches are open the meter should read 1 (for digital) or ∞ (for analogue). This will indicate that there is no connection between phase and neutral (short circuit).
Single-phase electricity – single-phase alternating current electricity is supplied to domestic dwellings and light commercial buildings at 230 volts 50 Hz. The supply is taken as a single tapping from a 415 volt three-phase four-conductor mains cable. Individual dwellings are connected to alternative phase conductors to balance out the load.
Single pole switch – a type of switch which is only connected to the phase conductor and is only capable of opening or closing the circuit. It does not isolate the circuit from the supply, therefore a danger exists from 'backfeed' down the neutral conductor. Isolation is only possible at the consumer unit. This type of switch is used for turning light fittings on and off or switching a boiler on when a thermostat calls for heat.
Three-phase electricity – in the UK electricity is generated as three-phase alternating current electricity. That is to say that three independent single phases are produced 120° out of step with each other as the generator rotates. The generator has three separate fields set 120° apart. The conventional method of identifying the three phases is by colour code; the three colours used are red, yellow and blue.
Transformer – a device used to either increase or decrease the line voltage to satisfy utilization; e.g. on building sites the mains voltage is stepped down from 230 volts to 110 volts in view of safety. The process is fairly simple: a laminated core is constructed from steel plates bolted together. Wound around the core is an incoming coil of insulated wire and an outgoing coil on the opposite side. One coil has more windings than the other. As the current flows through the first coil a magnetic flux is induced in the core, which flows around the core. The flow of flux induces a current in the second coil. The relationship of the coil sizes determines the increase or decrease of the secondary voltage. The following relationships apply to the electrical conversions that take place within a transformer when electricity flows:

$$\frac{\text{primary voltage}}{\text{secondary voltage}} = \frac{\text{primary turns}}{\text{secondary turns}} = \frac{\text{secondary amps}}{\text{primary amps}}$$

$$\frac{V_1}{V_2} = \frac{N_1}{N_2} = \frac{A_2}{A_1}$$

Example 7.6 A transformer is plugged into a 230 V supply for use with 110 V power tools. The turns of the primary coil on the transformer core are 100. How many turns of the secondary coil will produce 110 V?

$$\frac{V_1}{V_2} = \frac{N_1}{N_2} = \frac{230}{110} = \frac{100}{N_2}$$

$$N_2 = \frac{100 \times 110}{230} = \underline{\textbf{48 turns}}$$

If the transformer above is drawing 3 amps on the primary side, what current will be available for the power tools?

$$\frac{N_1}{N_2} = \frac{100}{48} = \frac{A_2}{3}$$

$$A_2 = \frac{100 \times 3}{48} = \underline{\textbf{6.25 amps}}$$

Volt (V) – the unit used to measure the force (pressure) which makes electrical current flow along conductors. An analogy might be the water stored in a cistern: when a valve is opened the water would flow, just as in the case of an electrical current being switched on. The voltage will cause the electrical current to flow.
Watt – the unit of electrical power. Electrical power is in fact a form of energy, therefore energy is measured in watts. One watt is equal to one joule per second. This can be expressed as work done per unit of time:

$$1 \text{ watt} = \frac{1 \text{ Joule}}{\text{second}} = \text{W} = \text{J/s}$$

The watt also has a relationship with electromotive force and current, whereby watts are equal to volts multiplied by amps. Because a watt is a relatively small amount of energy it is common to deal in kilowatts (1000 W).

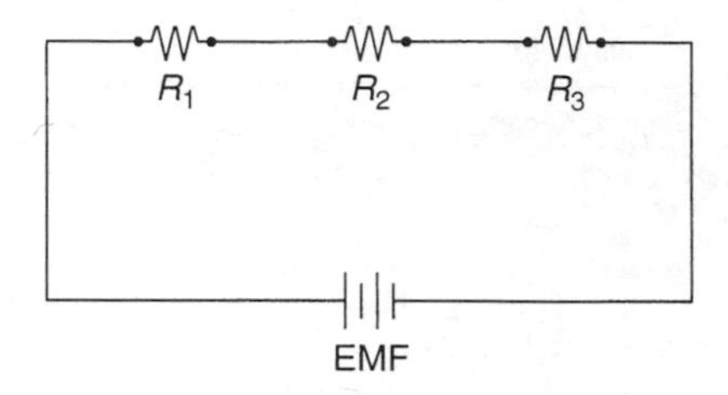

Fig. 7.1 Series circuit

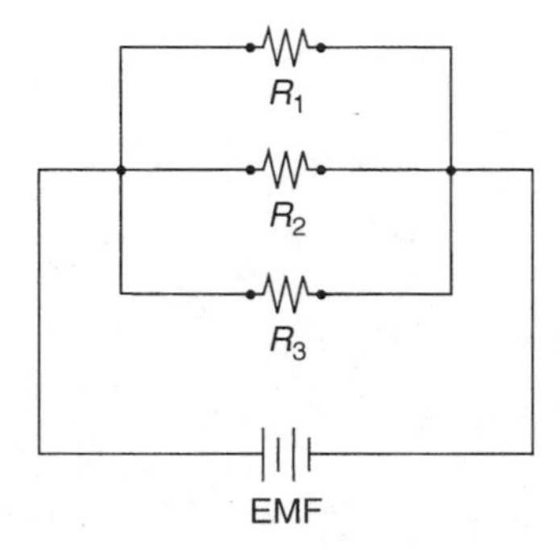

Fig. 7.2 Parallel circuit

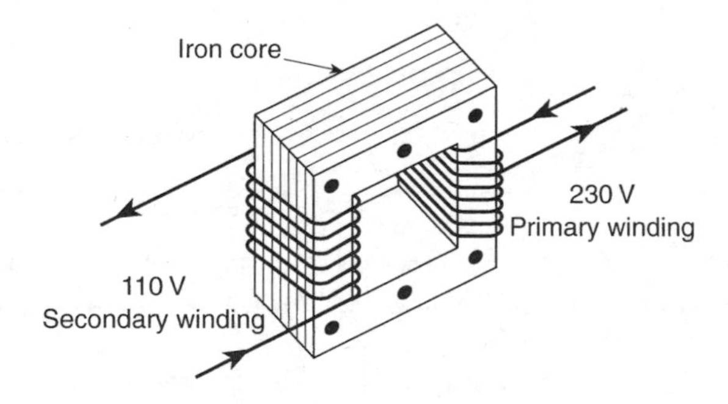

Fig. 7.3 Transformer

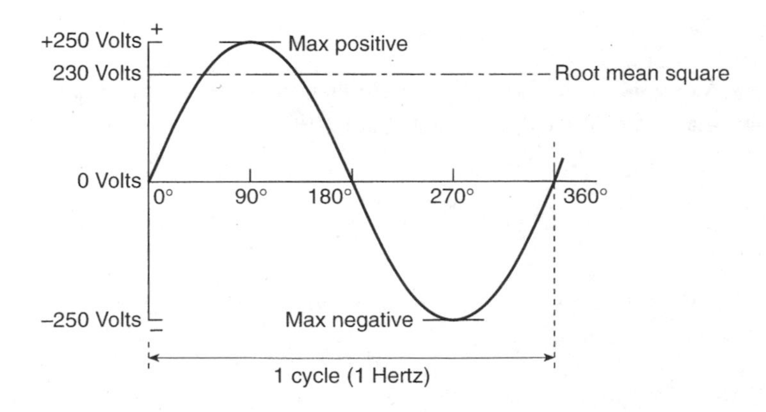

Fig. 7.4 Single-phase alternating current

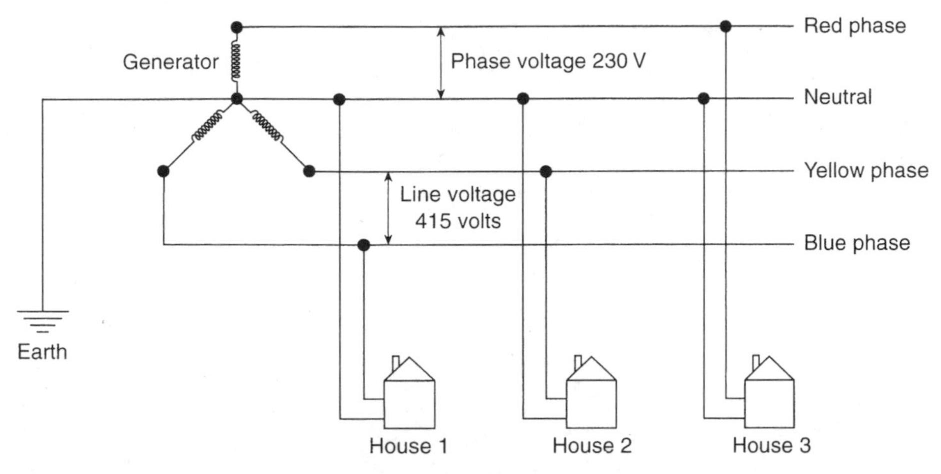

Fig. 7.5 Four-conductor three-phase distribution

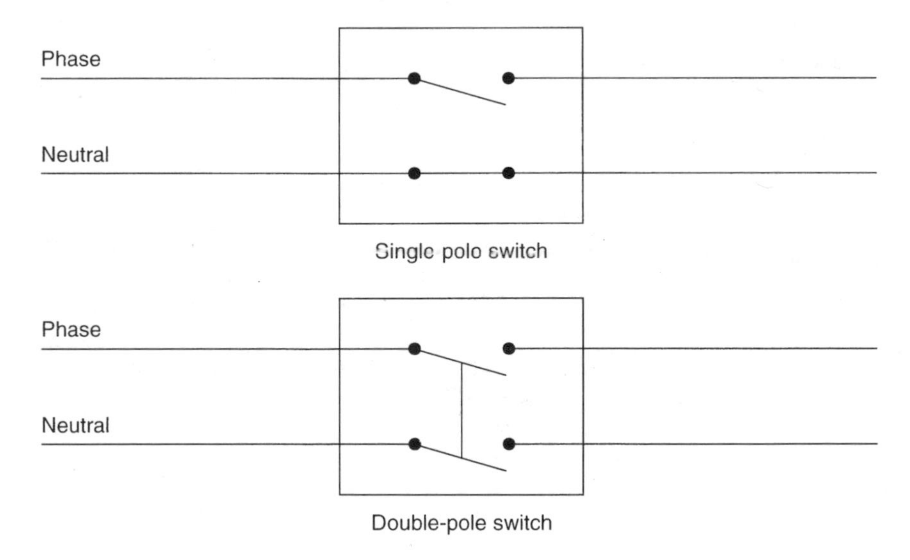

Fig. 7.6 Electrical switches

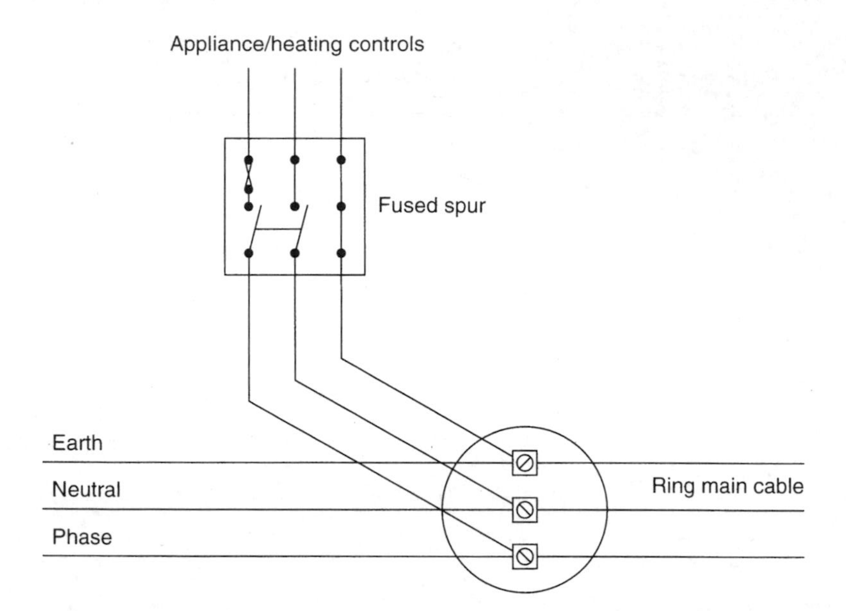

Fig. 7.7 Fused spur connection to ring main

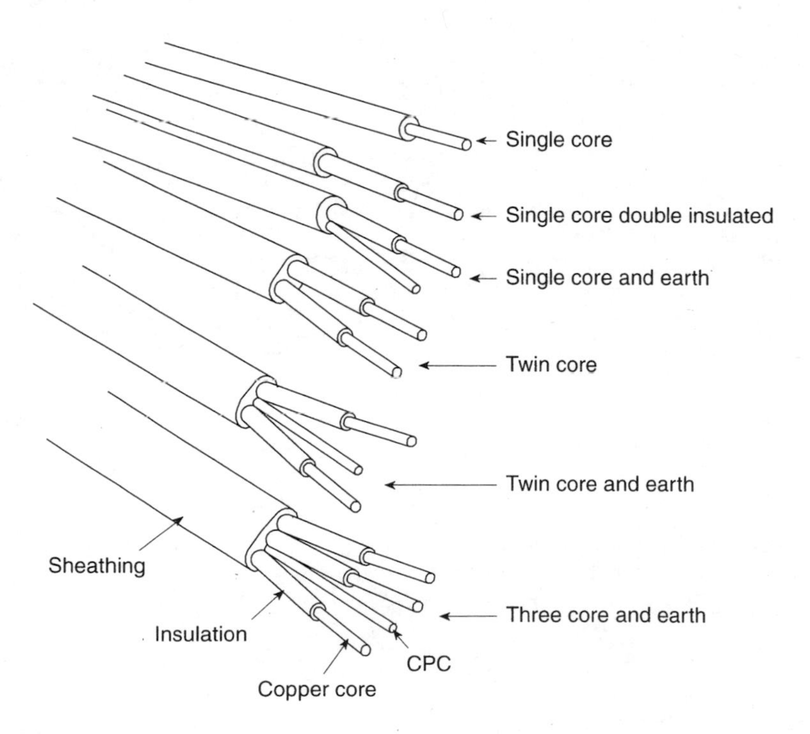

Fig. 7.8 PVC insulated cable

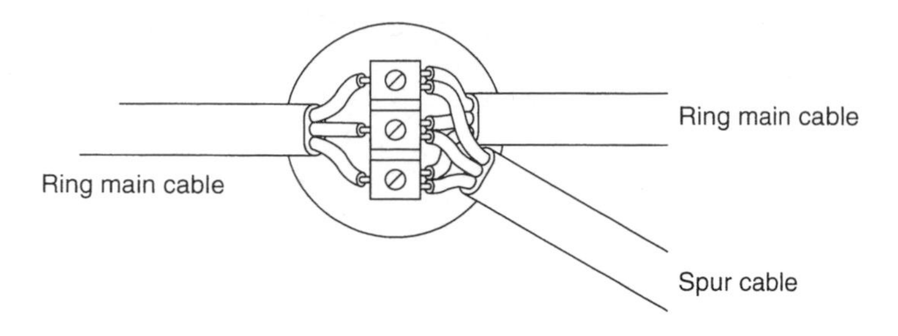

Fig. 7.9 Junction box

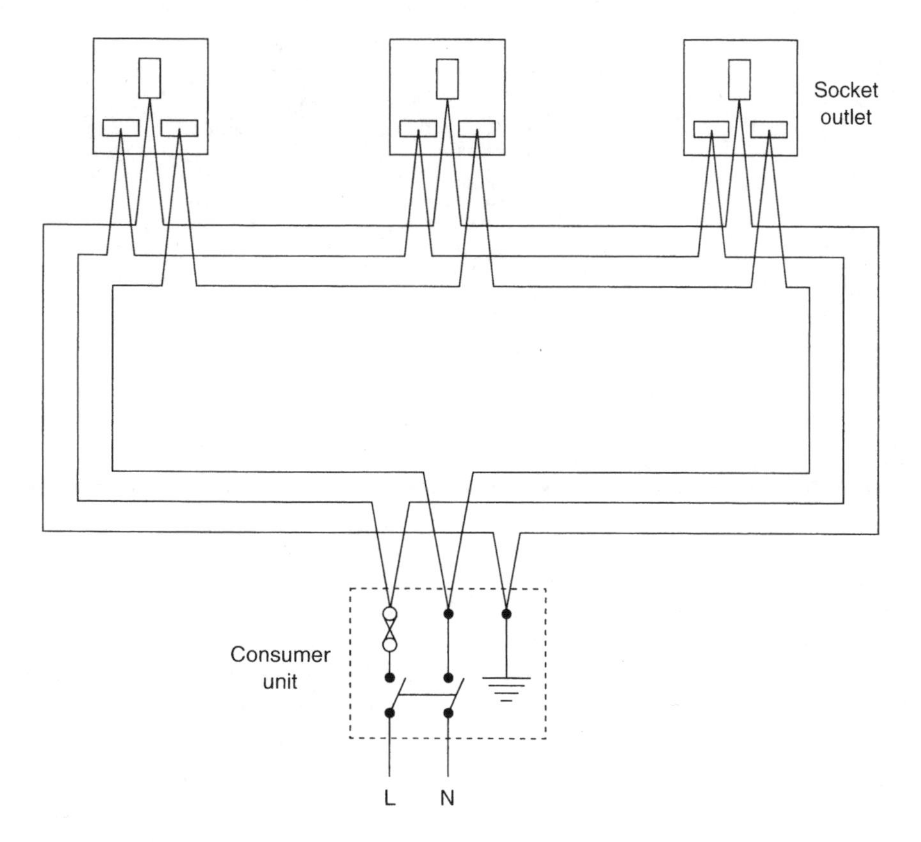

Fig. 7.10 Ring main circuit

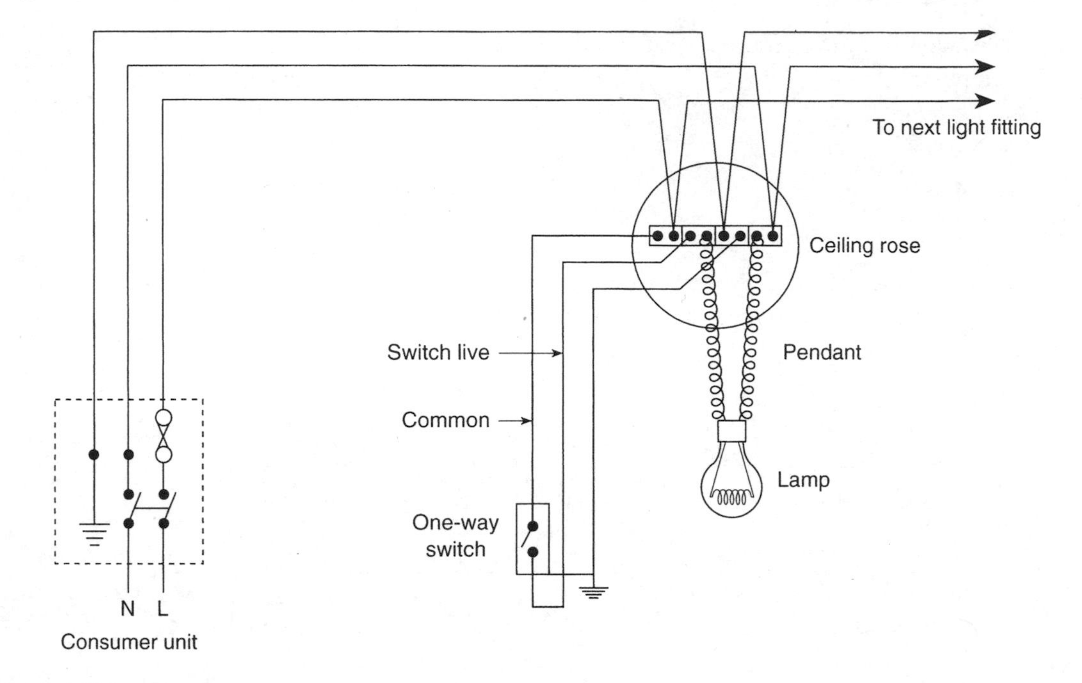

Fig. 7.11 Loop-in light circuit

Now try and answer these questions

1. A series circuit has the following resistors connected: 20 Ω, 30 Ω and 10 Ω. The total resistance of the circuit will be

   a 55 Ω
   b 65 Ω
   c 60 Ω
   d 50 Ω.

2. The following resistors are connected in series: 4.25 Ω, 2.62 Ω, 1.65 Ω and 5.46 Ω. The total circuit resistance will be

   a 12.25 Ω
   b 13.98 Ω
   c 15.25 Ω
   d 14.98 Ω.

3. There are two resistors of 20 Ω and 5 Ω connected in parallel to a circuit. What is the total resistance of the circuit?

   a 2.5 Ω
   b 5.2 Ω
   c 6 Ω
   d 4 Ω.

4. The following resistors are connected in parallel: 2.5 Ω, 4.25 Ω and 3.15 Ω. What is the total resistance of the circuit?

   a 1.04 Ω
   b 10.4 Ω
   c 9.9 Ω
   d 10.9 Ω.

5. The primary winding of a transformer has 50 turns and the secondary winding has 20 turns. If the primary voltage is 230 V, the secondary voltage will be

   a 92 volts
   b 460 volts
   c 52 volts
   d 250 volts.

6. Rewirable fuses in consumer units must comply to
   - a BS 6700
   - b BS 5440
   - c BS 3060
   - d BS 6644.

7. Cartridge fuses used to protect electrical installations are constructed to
   - a BS 1066; BS 1812; BS 66
   - b BS 1361; BS 1362; BS 88
   - c BS 1914; BS 1939; BS 47
   - d BS 2000; BS 1999; BS 46.

8. The conductor intended to protect electrical installations is referred to as the
   - a switch live conductor
   - b phase conductor
   - c neutral conductor
   - d continuous protective conductor.

9. Electrical current is measured in units referred to as
   - a amperes
   - b watts
   - c ohms
   - d volts.

10. The type of electricity supplied in the UK is referred to as
    - a alternating voltage
    - b alternating current
    - c direct current
    - d direct voltage.

11. Electricity generated in power stations is AC with a frequency of
    - a 100 cycles per second
    - b 200 cycles per second
    - c 25 cycles per second
    - d 50 cycles per second.

12. The system of overhead cables used to distribute electricity throughout the UK is called the
    - a Internet
    - b National Grid
    - c Electricity Network
    - d Transmission Web.

13. The type of electricity derived from batteries is called
    - a alternating voltage
    - b alternating current
    - c direct current
    - d direct voltage.

14. The term given to circuits providing electrical power in buildings is referred to as
    - a final circuits
    - b secondary circuits
    - c internal circuits
    - d mains circuits.

15. The phase conductor in fixed cable installations is identifiable by the insulation being coloured
    - a black
    - b brown
    - c yellow and green
    - d red.

16. The neutral conductor in fixed cable installations is identifiable by the insulation being coloured
    - a black
    - b brown
    - c yellow and green
    - d red.

17. The relationship between current, resistance and voltage is stated in
    - a Boyle's law
    - b Ohm's law
    - c Charges law
    - d BS 7671.

18. The statutory regulations which covers the use of electricity in the workplace are the

a *Electricity Supply Regulations*
b *Electricity at Work Regulations*
c *Electrical Appliance Regulations*
d *Electricity Low Voltage Regulations.*

19. The regulations designed to protect persons, property and livestock against hazards arising from electrical installations are the

a *Electricity at Work Regulations*
b *Electrical Appliances Regulations*
c *IOP Design Guide*
d *16th Edition IEE Wiring Regulations.*

20. The type of electrical circuit used to connect a single appliance to the consumer unit is a

a ring main
b fused spur
c radial circuit
d loop-in circuit.

21. An electrical component which is capable of converting an alternating current to a direct current is referred to as a

a rectifier
b capacitor
c resistor
d transformer.

22. The electrical unit used to measure electromotive force is

a amperes
b watts
c ohms
d volts.

23. The conductor in a final circuit which is protected by a circuit protective device is the

a earth conductor
b phase conductor
c neutral conductor
d continuous protective conductor.

24. Single-phase electricity is supplied as

a 415 V; AC; 50 Hz
b 230 V; AC; 50 Hz
c 415 V; DC; 100 Hz
d 230 V; DC; 100 Hz.

25. The supplier's main cable is connected to the internal wiring by means of a device referred to as a

a service cut-out device
b residual current device
c terminal block
d junction box.

26. A ring main power circuit is protected by a circuit protective device rated at

a 5 amps
b 15 amps
c 20 amps
d 30 amps.

27. When connecting a spur cable to a ring main final circuit the installation fitting used is referred to as a

a consumer unit
b ceiling rose
c junction box
d socket box.

28. The device used to connect the equipotential bonding conductor to metallic service pipes is known as a

a earth clamp
b terminal block
c link
d pipe clamp.

29. Resistance to electrical flow in a conductor is measured in units known as

a amperes
b watts
c ohms
d volts.

30. The unit of electrical power is measured in

a amperes
b watts
c ohms
d volts.

31. Electricity generated in the UK is three-phase electricity. In the generator the three phases are set

a 90° apart
b 45° apart
c 180° apart
d 120° apart.

32. A system of electrical installation using tubing and fittings with single cables is referred to as a

a small-bore system
b pipework system
c conduit system
d pyro system.

33. A type of switch capable of disconnecting both the phase and neutral conductors of a circuit is known as a

a double-pole switch
b two-way switch
c circuit breaker
d single-pole switch.

34. A double-pole switch is recommended for isolating __________ circuit.

a a lighting
b a bonding
c an immersion heater
d an emergency lighting.

35. The system that prevents fixed metalwork in buildings developing a potential difference (PD) is known as

a continuous protective conductor
b lightning conductor
c earth leakage
d equipotential bonding.

36. The three colours used to identify conductors in flexible cables are

a brown, black, yellow/green
b brown, blue, yellow/green
c red, blue, yellow
d red, black, green.

37. If a permanently wired appliance is to be connected to a ring main a ______ must be installed.

a fused spur
b single-pole switch
c connector block
d taped joint.

38. The final circuit associated with lighting installations is known as a

a ring main
b loop-in circuit
c radial circuit
d single-wire circuit.

39. A micro-circuit breaker is a mechanical tripping mechanism which works on the

a Bourdon principle
b expansion of liquid principle
c gravity principle
d bimetallic principle.

40. One advantage of using micro-circuit breakers to protect final circuits is that they are

a cheap
b easy to override
c resettable
d made of plastic.

41. The component that enables all final circuits to be terminated, protected and isolated is referred to as a

a power unit
b consumer unit
c disposal unit
d service cut-out.

42. The reason for carrying out a continuity test is to ensure there is a direct path to

 a earth
 b the fuse
 c the switch
 d flow.

43. The purpose for carrying out a polarity test is to ensure that the phase conductor is connected to the

 a appliance live terminal
 b appliance earth terminal
 c appliance neutral terminal
 d appliance casing.

44. The type of electrical switch that will only disconnect the phase conductor in an electrical circuit is known as a

 a double-pole switch
 b two-way switch
 c circuit breaker
 d single-pole switch.

45. An immersion heater has a 2500 watt rating at 250 volts. What size circuit protective device is required to protect it?

 a 30 amps
 b 5 amps
 c 10 amps
 d 20 amps.

Now check your answers from the grid.

Q 1; c	Q 10; b	Q 19; d	Q 28; a	Q 37; a
Q 2; b	Q 11; d	Q 20; c	Q 29; c	Q 38; b
Q 3; d	Q 12; b	Q 21; a	Q 30; b	Q 39; d
Q 4; a	Q 13; c	Q 22; d	Q 31; d	Q 40; c
Q 5; a	Q 14; a	Q 23; b	Q 32; c	Q 41; b
Q 6; c	Q 15; d	Q 24; b	Q 33; a	Q 42; a
Q 7; b	Q 16; a	Q 25; a	Q 34; c	Q 43; a
Q 8; d	Q 17; b	Q 26; d	Q 35; d	Q 44; d
Q 9; a	Q 18; b	Q 27; c	Q 36; b	Q 45; c

8

COLD WATER SUPPLY

To tackle this chapter you will need to know:

- cold water distribution;
- direct cold water supply systems;
- indirect cold water systems;
- faults in water systems;
- *Water Bylaws*.

GLOSSARY OF TERMS

Ball valve – a float-operated valve used as an automatic device which responds to the water level within the cistern, opening when water is drawn off and closing when the required water level is restored. This device is designed to prevent wastage of water.

Boundary stop valve – a valve placed outside the boundary of the property which it serves. It is the property of the water supplier and is a means of turning off the supply to the building. It provides isolation for the repair of the service pipe.

Branch pipe – a pipe which connects the sanitary appliances and the rising main.

Bylaw 30 – Bylaw 30 of the *Model Water Bylaws* describes the requirements for stored cold water in buildings, the requirements of which state that the cistern must be covered by an air-tight lid with a screened vent and overflow connection to prevent vermin and bugs entering the stored water.

Cold water storage cistern – a storage vessel for holding a predetermined amount of water for use in an indirect cold water supply system or/and a vented hot water supply system.

Communication pipe – a short length of pipe connecting the boundary stop valve and the ferrule on the water main.

Direct cold water supply – in this type of system mains cold water is available at all draw-off points. Cold water is not stored for drinking purposes. The pipework is usually of small diameter – 15 mm.

Distribution pipe – the main feed pipe running from the cold water storage cistern to connect all the branch pipes to the stored water.

Drain valve (DOC) – a small device which is intended for the purpose of draining the isolated system only. It should be situated directly above the service stop valve.

Equilibrium ball valve – a special type of ball valve which through its structure and configuration has the mains pressure equal on both ends of the valve piston. This is achieved by means of a hole passing through the piston, and the piston chamber being sealed.

Ferrule – a type of valve which has a construction which enables it to be rotated through 360°. In the case of a cast iron main, the ferrule is of brass and can be placed into the main without turning off the main by means of a sealed chamber which is strapped to the main. Likewise with plastic mains the ferrule can be fused to the main before the incision is made. The ferrule has a plug-type mechanism and is normally buried along with the main.

Indirect cold water supply – in this type of system cold water is stored in a cold water storage cistern at high levels. In some cases a single tap is connected directly to the mains – usually the kitchen sink tap. Since the introduction of 'Bylaw 30' potable water may now be stored to serve all cold water taps. This system provides a limited supply if the water main is turned off.

Isolation valve – typically a gate valve whose function is to isolate the stored cold water from the hot or indirect cold water systems for maintenance and repair.

Oscillating ball valve – due to extremely high water pressures some ball valves, or float-operated valves, will start to oscillate when rapid discharge from the ball valve causes turbulence of the water surface which in turn makes the float jump up and down violently. This has a similar effect to water hammer. It can be remedied by either fitting a pneumatic bottle to the rising main or fitting an equilibrium ball valve.

Pneumatic bottle – a sealed section of pipe larger in diameter than the rising main, usually two diameter sizes up, approximately 150 mm long. It has the effect of providing an air cushion trapped within it which is compressed and expanded as the float moves up and down.

Rising main – the pipe which connects the service stop tap and cold water storage cistern. All other pipes connected to it are referred to as branch pipes.

Service pipe – the pipe which is laid underground and connects the internal pipework to the boundary stop tap. It is recommended that the service pipe enters the building on an internal wall. The service pipe should have a minimum ground cover of 750 mm.

Service valve – a requirement of the *Model Water Bylaws* wherever a float-operated valve (ball valve) is fitted. Its purpose is for maintenance and repair. It is normally a quarter turn 'ball'-type valve. It should be noted that in Scotland the Water Authorities favour the use of other valves for this purpose due to higher pressures. The rapid action of the quarter turn valve would cause water hammer.
Service stop valve – a valve situated within the first 150 mm of the service pipe where it enters the building and is the occupier's means of isolating the supply to the building.
Velocity noise – a problem normally associated with 'direct' cold water systems and is the result of high pressure within the pipework causing the water to flow extremely fast, the result being audible noise emanating from the pipework.
Water hammer – a problem which is sometimes associated with older systems. It is caused when water fittings, particularly float-operated valves, become worn or if pipework has not been correctly clipped to the structure. As the name suggests this problem makes a noise like someone 'hammering' the pipes.
Water main – part of an underground network of distribution pipes which supply water to the districts of towns and villages. It is the property of the supplier who is also responsible for its upkeep.

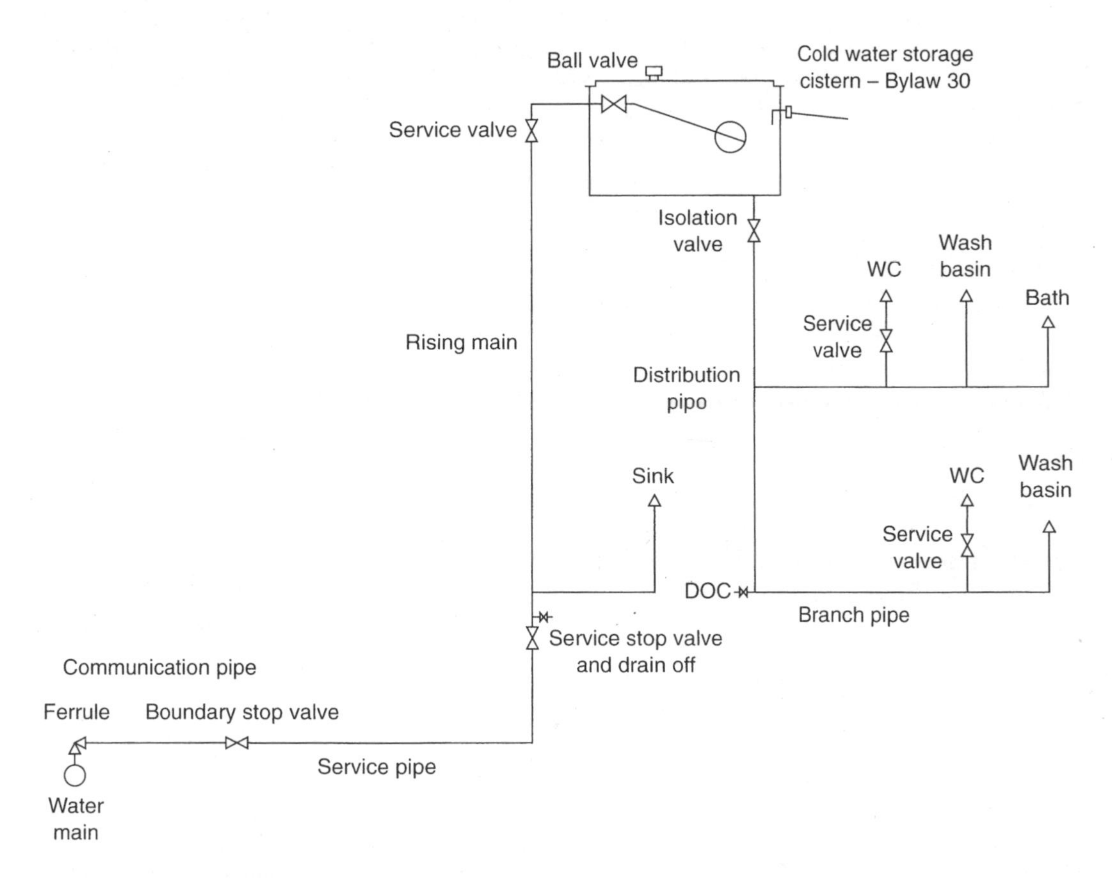

Fig. 8.1 Indirect cold water system

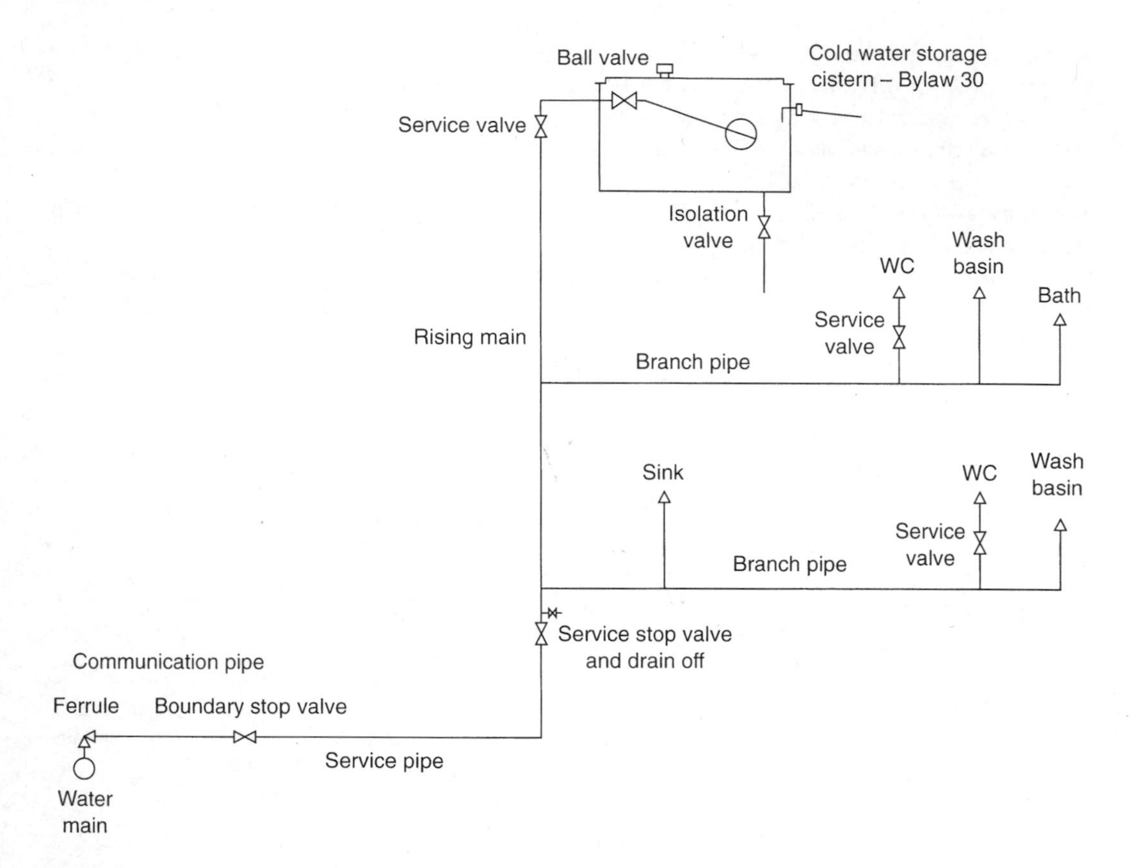

Fig. 8.2 Direct cold water system

Now try and answer these questions

1. The system of pipework which is to be found buried beneath the roads and streets which is used to convey water is referred to as the

   a water mains
   b service mains
   c aquifer
   d sewer pipe.

2. Water mains are the property of the

   a householder
   b Local Authority
   c Water Authority
   d Area Health Trust.

3. Water mains are constructed of either

   a lead or copper
   b asbestos or low-carbon steel
   c cast iron or asbestos
   d cast iron or polyethylene.

4. An insertion or connection to a water main is made by placing a special type of valve directly on top of the main known as a

   a ferrule
   b stop valve
   c gate valve
   d main valve.

5. The special property of the ferrule is that it can be inserted into the water main using a clamp-on chamber

   a without having to dig holes in the ground
   b without having to turn off and drain down the water main
   c after turning off the water main and draining down
   d having first informed the people in the street.

6. The ferrule is a means of

   a preventing water being wasted from the main
   b providing the most economical connection
   c isolating the supply when the customer requires it
   d isolating and disconnecting the service to a building.

7. The short length of buried pipe connecting the ferrule and the boundary stop valve is referred to as the

   a underground pipe
   b service pipe
   c communication pipe
   d submain pipe.

8. The communication pipe is wholly the property of the

   a building owner
   b Water Authority
   c local council
   d Area Health Authority.

9. The boundary stop valve is normally found

   a just outside the perimeter of the property which it feeds
   b just inside the perimeter of the property which it feeds
   c in the road outside the property which it feeds
   d outside the front door.

10. To enable access to the boundary stop valve, it is normally found contained within a

   a brick-built chamber with a hinged lid
   b wall-mounted box with a locked door
   c chamber made of a section of drain pipe with a hinged lid
   d container buried in the ground.

11. The boundary stop valve is wholly the property of the

   a local council
   b building owner
   c Water Authority
   d Area Health Authority.

12. From the boundary stop valve to the building is a length of pipe referred to as the

   a rising main
   b communication pipe
   c service pipe
   d main branch pipe.

13. The service pipe must be buried under the ground level to a minimum recommended depth of

   a 750 mm
   b 900 mm
   c 650 mm
   d 700 mm.

14. The service pipe is buried beneath the ground to the recommended depth to prevent damage from

   a traffic passing over it and squashing the pipe
   b moles biting through it
   c rain water corroding it
   d frost penetrating down to it and freezing the pipe.

15. There are two common materials used for manufacturing pipes which are suitable for laying as service pipes. They are either

a lead or copper
b polyethylene or copper
c polyethylene or lead
d low-carbon steel or copper.

16. The *Water Bylaws* suggest that the service pipe should be brought into the property on

a one of the outside walls
b the sink unit cold connection
c to the wall nearest the kitchen
d to an internal wall.

17. The service pipe terminates inside the building. At the end of the pipe there is a control device known as a

a service stop valve
b gate valve
c drain-off cock
d globe valve.

18. The service stop valve should be fitted no more than ________ above the floor.

a 150 mm
b 1500 mm
c 400 mm
d 50 mm.

19. The device that is fitted immediately above a service stop valve is known as a

a stop valve
b gate valve
c drain-off valve
d globe valve.

20. The pipe which connects the service stop valve and the cold water storage cistern is referred to as the

a direct pipe
b main pipe
c connecting pipe
d rising main.

21. In the direct system of cold water supply all ________ are connected directly to the rising main.

a cold water draw-off points
b low-pressure draw-offs
c sanitary appliances draw-offs
d shower fittings.

22. In the indirect system of cold water supply only ________ are connected directly to the rising main.

a the WC cisterns
b sanitary appliances
c the cold water storage cistern and one tap
d shower fittings.

23. One advantage of the direct system of cold water supply is that

a the water is a lot cooler in summer
b mains pressure is available at every draw-off point
c there is more velocity noise
d a bath is easier to fill.

24. One advantage of the indirect system of cold water supply is that

a mains pressure is only available at one draw-off point
b there is more pipework involved
c stored water is more likely to become polluted
d water is still available if the water main is turned off.

25. One disadvantage of the direct system of cold water supply is that

a the water pressure can be excessive
b taps drip more often
c if the water mains are turned off there is no cold supply
d the water is icy cold in winter.

26. One disadvantage of the indirect system of cold water supply is that

a stored water is available at peak usage times
b it is more costly to install
c there is a great deal of velocity noise
d it has to be inspected annually.

27. The term given to the pipes which are connected between the rising main and the draw-off points is

a service pipes
b communication pipes
c connecting pipes
d branch pipes.

28. Where a branch pipe or rising main is connected to a float-operated valve the *Water Bylaws* recommend that a control device known as a _______ should be fitted.

a service valve
b fire valve
c sluice valve
d gate valve.

29. The purpose of a service valve is to allow access to the float-operated valve for

a preventing water hammer
b preventing back siphonage
c maintenance and repair
d preventing excessive use.

30. A float-operated valve is a device which allows water to flow into the storage cistern

a when someone turns it on
b when the water pressure builds up
c automatically
d when the water pressure falls.

31. A float-operated valve relies on the principle of _______ for its operation.

a expansion
b levers
c Pythagoras
d inertia.

32. A float-operated valve will respond to the rise and fall of the _______ in the storage cistern.

a temperature of the water
b density of the water
c pressure of the water
d water level.

33. The vessel used to hold water in storage is referred to as a

a cold water tank
b hot water storage cistern
c cold water storage cistern
d hot water tank.

34. Stored cold water in buildings for human consumption is covered in the *Model Water Bylaws* by

a Bylaw 30
b Bylaw 3
c Bylaw 13
d Bylaw 25.

35. The requirements of the *Water Bylaws* insist that a cistern must have

a a metal base
b a sealed lid and insect screens on the overflow and vent
c a non-removable lid
d the overflow connection above the cold inlet.

36. To prevent the contents of the cold water storage cistern being frozen during the winter months it is recommended that it is provided with

a several blankets placed over it
b a paraffin lamp to keep it warm
c a loose fitting insulating jacket
d a good-quality insulation jacket.

37. The *British Standards Code of Practice* BS 6700 recommends that in a roof space the loft insulation should be

a doubled to prevent the water from heating up
b removed directly underneath the cistern to keep it from freezing
c completely wrapped round the cistern
d of the highest-quality Rockwool.

38. The term given to the pipes which are connected from a storage cistern to the draw-off points is

a service pipes
b communication pipes
c distributing pipes
d branch pipes.

39. When the cold water pipework of a system starts to make loud intermittent banging sounds this fault is referred to as

a water hammer
b noisy neighbours
c hard water flowing in the pipes
d bits of scale entrained from the water main.

40. One of the faults to be found in cold water systems which is called 'water hammer'can be an indication that a ______ is worn.

a section of pipework
b service valve
c service stop valve
d float-operated valve.

41. Direct cold water systems in areas with very high mains pressure are prone to giving off noise when water flows. This is referred to as

a rapid draw-off
b velocity noise
c excessive pressure
d partial blockage.

42. When water discharges from a float-operated valve into a cistern it may cause turbulence of the water surface. This will often result in noise coming from the pipework. This is usually caused by

a loose pipe clips
b defective service valve
c oscillating ball valve
d blocked overflow.

43. One of the remedies for the fault in Question 42 might be to install a

a pneumatic bottle
b new ball valve
c extra pipe clips
d repipe of the system.

44. The working principle of the 'pneumatic bottle' is to have a cushion of ____ trapped within the bottle.

a compressed air
b air
c nitrogen
d hydrogen.

45. To overcome the problems created by high-pressure mains in storage cisterns the float-operated valve is replaced with a

a torbeck valve
b class 2 ball valve
c equilibrium ball valve
d pillar tap.

Now check your answers from the grid.

Q 1; a	Q 10; c	Q 19; c	Q 28; a	Q 37; b
Q 2; c	Q 11; c	Q 20; d	Q 29; c	Q 38; c
Q 3; d	Q 12; c	Q 21; a	Q 30; c	Q 39; a
Q 4; a	Q 13; a	Q 22; c	Q 31; b	Q 40; d
Q 5; b	Q 14; d	Q 23; b	Q 32; d	Q 41; b
Q 6; d	Q 15; b	Q 24; d	Q 33; c	Q 42; c
Q 7; c	Q 16; d	Q 25; c	Q 34; a	Q 43; a
Q 8; b	Q 17; a	Q 26; b	Q 35; b	Q 44; b
Q 9; a	Q 18; a	Q 27; d	Q 36; d	Q 45; c

9

DOMESTIC HOT WATER SUPPLY

To tackle this chapter you will need to know:

- the direct hot water storage system;
- the indirect hot water storage system;
- the use of secondary return systems;
- the installation of unvented stored hot water systems;
- the various components found on DHW systems.

GLOSSARY OF TERMS

Automatic air vent (AAV) – a component designed specifically to automatically purge air from pipework as and when it occurs in the system. It is normally positioned at high points on the system.
Bath – a sanitary appliance for personal hygiene.
Boiler – a fuel-burning appliance with a heat exchanger as part of or within the combustion chamber. For direct hot water systems the heat exchanger is normally constructed from copper.
Cold feed and expansion pipe – the pipe which connects a feed and expansion cistern to the primary side of a double-feed indirect hot water system. Its function is to top up the primary side and enable expanded water to return to the cistern.
Cold feed pipe – a pipe that connects the cold water storage cistern to the hot water storage cylinder. Its function is to replace hot water drawn off from taps with the stored cold water. Its connection to the cistern should provide a 25 mm deep catchment for sediments to collect.
Cold water storage cistern – a storage vessel for holding a predetermined amount of water for use in an indirect cold water supply system or/and a vented hot water supply system.
Dead-leg – a pipe that does not form part of a circuit – the water is static when the taps are closed. BS 6700 gives recommendations for maximum lengths of dead-legs. The limitation of dead-leg length is designed to minimize wastage of water and maximize energy consumption.
Direct cylinder – a copper storage vessel which normally has four connections for pipework and one for an electric immersion heater. Heated water from the boiler flows directly into the mass of stored water and is constantly recycled until the water reaches the boiler water temperature.
Drain valve (DOC) – a small device which is intended for the purpose of draining the isolated system only. It should be situated directly above the service stop valve.
Expansion vessel – a component associated with sealed systems, i.e. systems not open to atmospheric pressure. The vessel is divided into two spaces separated by a flexible bladder; one space is connected to the waterway of the system and the other is filled with air at a predetermined pressure. As the system expands the increase in volume fills the bladder and thus compresses the air in the air space. On commissioning or maintenance this pressure is set at the manufacturer's recommended pressure. Therefore the water expansion volume occupies the majority of the air space.
Feed and expansion cistern – a component which has several functions within the system: firstly, to fill the system with water; secondly, to automatically top up any water lost through leaks, etc.; thirdly, to accommodate the volume of expanded water upon heating; and fourthly, to accommodate any additional expansion due to failure of automatic controls.
Indirect cylinder – a storage vessel which keeps separate the stored hot water from the primary heating water. This is achieved by means of a heat exchanger either of the coil type or the annulus pattern. It enables a cast iron or steel boiler to be used for domestic water heating.
Indirect (unvented) cylinder – a specially designed and constructed vessel capable of withstanding greater pressures. Indirect cylinders are constructed from thick sheet copper or glass-lined steel vessels. They are normally supplied as a complete ready-to-fit package containing all the mandatory safety devices. They are also available as direct cylinders in complete units.
Non-return valve (NRV) – a component which is designed to permit water to flow in one direction only; any tendency for the water to reverse its flow is immediately checked and prevented from returning back towards the mains. This valve is intended to close when the heated water expands, therefore forcing any expansion volume into the expansion vessel.

Open vent pipe – a connection to the primary or secondary flow pipe at its highest point from where it is taken to a point where it can discharge any water 'spouted' from the system over the feed cistern. This pipe is open to atmospheric pressure and therefore prevents the water boiling point from going higher than 100°C. It also permits air to escape automatically from the system and in addition prevents siphonic effects developing, which could under certain circumstances collapse the cylinder.
Pipework – traditionally, these systems are constructed using copper tube to BS 2871; Table X with capillary fittings containing lead-free solder. Alternatives to copper are galvanized low-carbon steel tube or one of the many plastic plumbing systems, e.g. HEP20.
Point of delivery – the point at which stored hot water enters the pipework system.
Point of discharge – the point where hot water flows from taps for use in sanitary appliances, etc.
Pressure reducing valve – a component designed to give a constant downstream water pressure regardless of fluctuating mains pressure. Mains pressure acting upon a flexible diaphragm is countered by a thrust spring which is tensioned to a predetermined value. The forces acting on the diaphragm will ultimately position a valve in proximity to its seating, thereby throttling the water. Constant pressure is a means of controlling the boiling point of the water.
Pressure-relief valve – a device which, although not strictly defined as a safety device, is used to maintain the pressure of the system constant. The pressure of the system acts upon a valve that is spring loaded to a predetermined pressure value. Should the water pressure increase above the relief pressure setting the valve will lift and water will flow to waste until the status quo is regained, whereupon the valve will reseat. This process limits the amount of water that flows to waste.
Primary flow pipe – a pipe that connects the top connection on the boiler heat exchanger to the top connection of the storage vessel. It is one-half of the first or 'primary' circuit within the system.
Primary return – the second half of the primary circuit connecting the bottom connection on the heat exchanger to the lower connection on the storage vessel.
Pump (circulator) – a pump to circulate a volume of hot water at storage temperature around the system. The type of pump used in secondary circulation systems should be a bronze-bodied type to prevent discoloration of the water. It is best situated on a 'bypass' connected to the secondary flow pipe close to the storage vessel. Normally the pump does not assist the flow of water from the taps.
Rising main – the pipe which connects the service stop tap and cold water storage cistern. All other pipes connected to it are referred to as branch pipes.
Secondary flow pipe – a pipe that connects the top of the storage vessel to the draw-off points (taps) on the systems. The diameter of the pipework will depend upon the number of taps served by the pipe.
Secondary return pipe – a pipe added to secondary distribution pipework when the distance between the 'point of delivery' and the 'point of discharge' exceeds those imposed by BS 6700. Its function is to permit the circulation of stored hot water through the secondary flow system to maintain the stored hot water temperature within the pipework in readiness for discharge. These pipes are normally of a smaller diameter as their function is to circulate sufficient water to compensate for heat loss from the secondary flow pipes.
Service stop valve – a valve situated within the first 150 mm of the service pipe where it enters the building and is the occupier's means of isolating the supply to the building.
Service valve – this valve is a requirement of the *Model Water Bylaws* wherever a float-operated valve (ball-valve) is fitted. Its purpose is for maintenance and repair. It is normally a quarter turn 'ball'-type valve.
Sink – a sanitary appliance normally situated in a kitchen or utility room.
Thermal relief valve – a temperature-operated relief valve inserted into the top of an unvented storage cylinder. It is set to open if the stored water reaches 95°C. A spring-loaded valve is opened by the expansion of a temperature-sensitive fluid contained in a phial. The expanding fluid pushes a plunger against the valve and thus opens it allowing water to flow to waste. As the water temperature falls the fluid contracts and the valve closes again.
Tun-dish –a receptacle for discharged water from a relief valve to enter the waste pipe. It also provides an air gap thus preventing siphonic tendencies.
Wash basin – a sanitary appliance found in bathrooms and bedrooms.

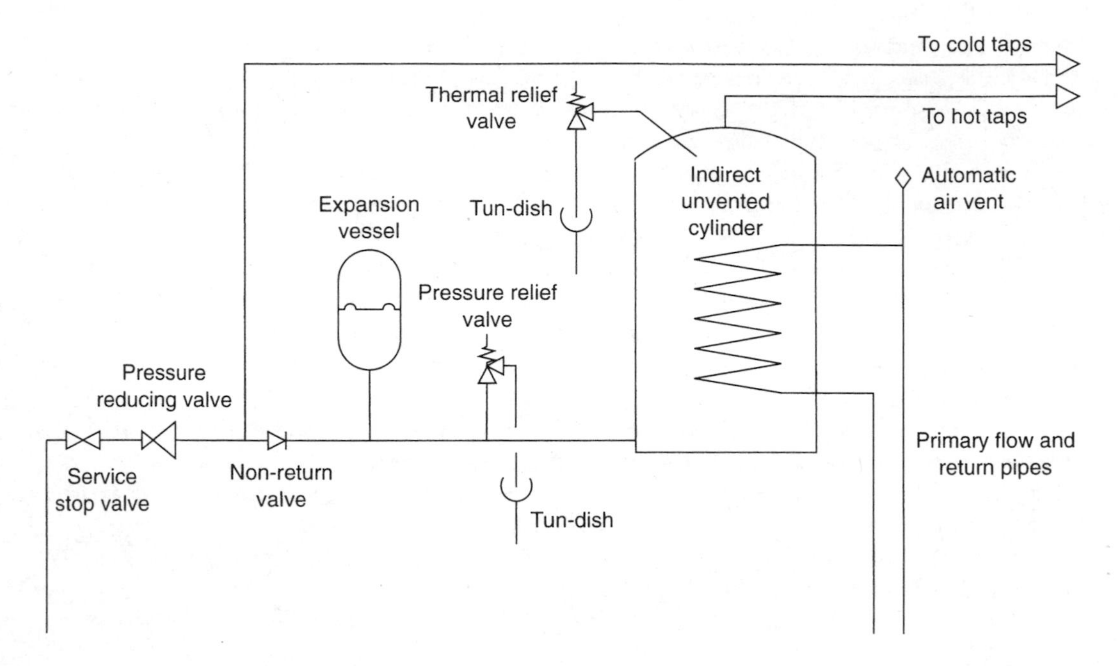

Fig. 9.1 Unvented hot water supply

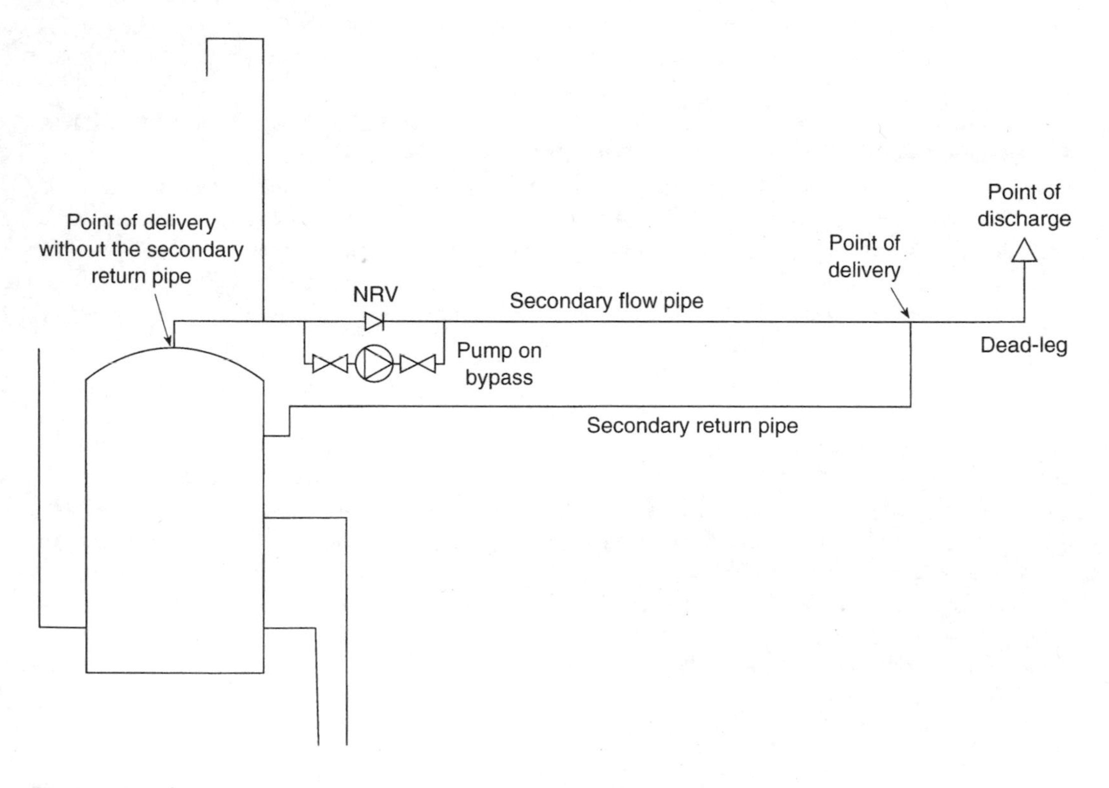

Fig. 9.2 Secondary return system

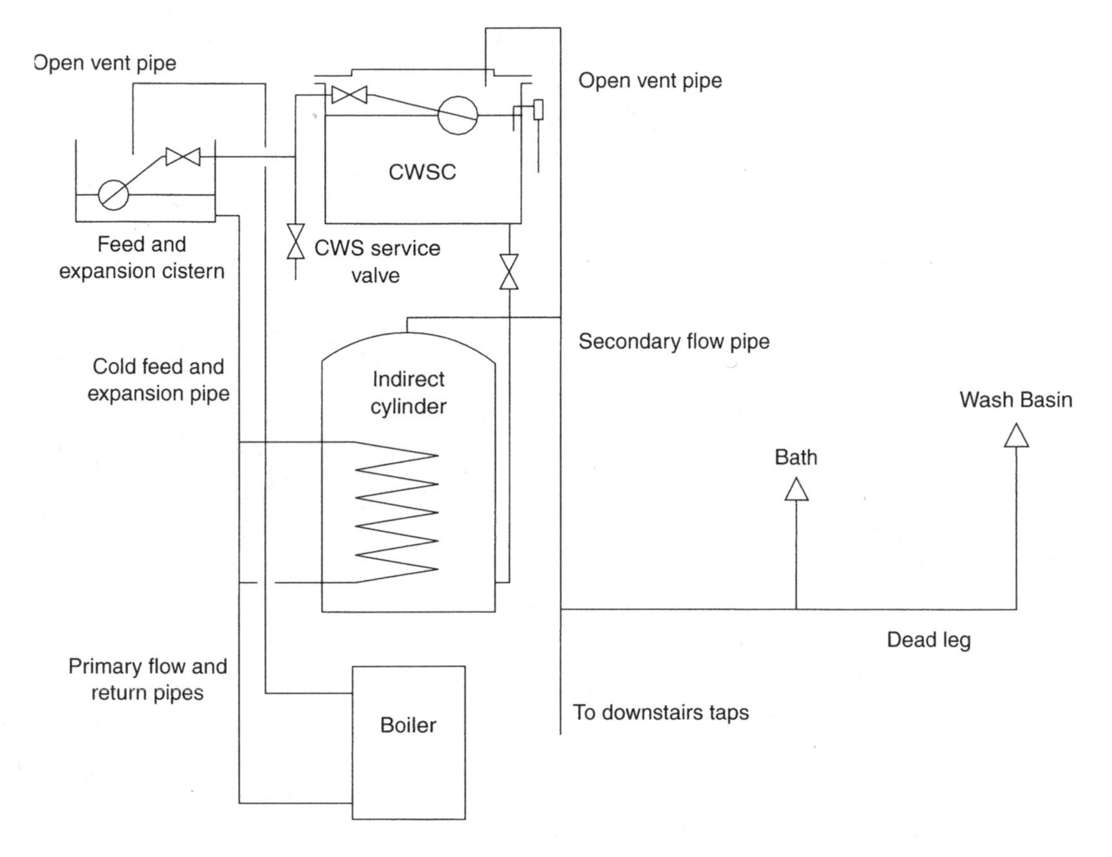

Fig. 9.3 Indirect hot water supply

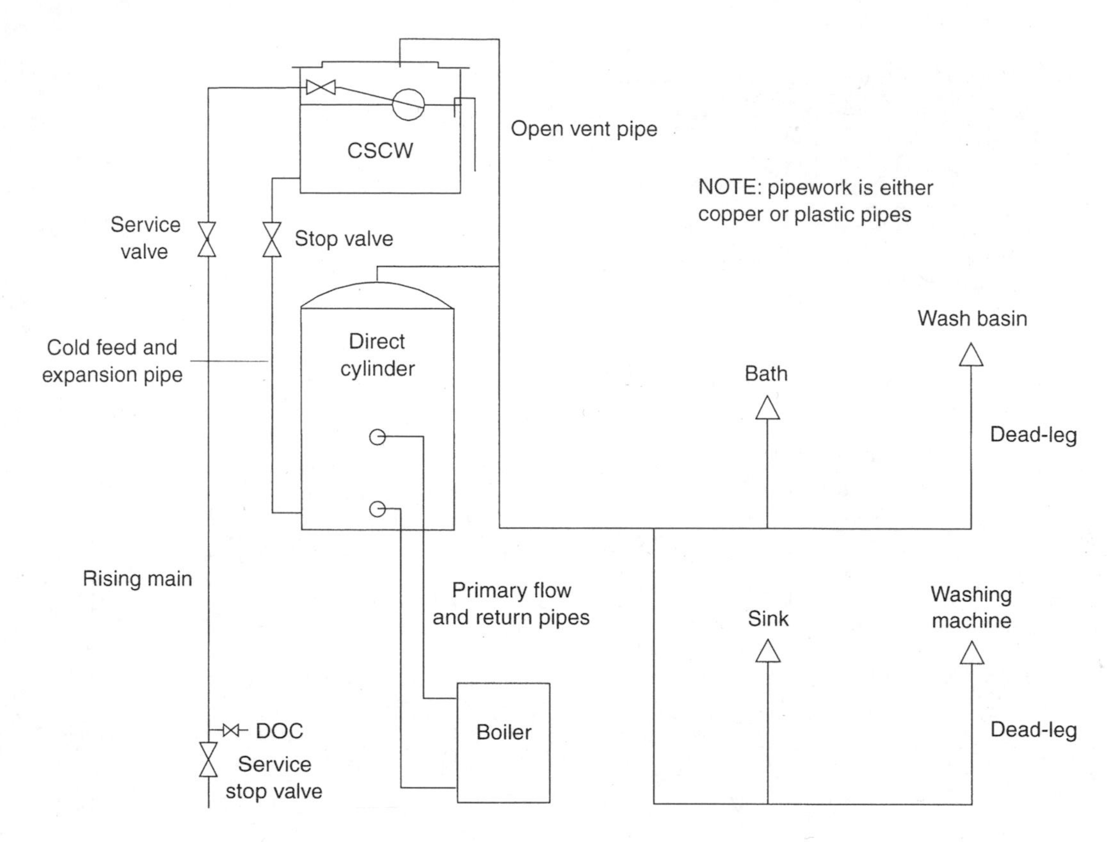

Fig. 9.4 Direct hot water supply

Now try and answer these questions

1. A component designed to remove air from wet systems without the use of a release key is known as
   a an air release valve
   b an open-vent pipe
   c an expansion vessel
   d an automatic air vent.

2. The most appropriate position to fix an air release device is at the
   a highest part of a system
   b point nearest the cylinder
   c lowest part of the system
   d adjacent to a tun-dish.

3. The pipe connecting a storage vessel or boiler to a feed and expansion cistern is referred to as the
   a open-vent pipe
   b primary return pipe
   c cold feed and expansion pipe
   d secondary flow pipe.

4. A fuel-burning appliance used to heat up stored water is commonly referred to as a
   a combi-boiler
   b instantaneous water heater
   c boiler
   d multi-point heater.

5. With respect to a direct hot water system the heat exchanger in the boiler should be manufactured from

- **a** cast iron
- **b** glass
- **c** low-carbon steel
- **d** copper.

6. The main function of the cold feed and expansion pipe is to permit

- **a** the expanding volume of water to pass to the feed and expansion cistern
- **b** air to escape from the system and prevent airlocks
- **c** siphonic action occurring in the secondary flow pipe
- **d** the control of the water pressure.

7. It is permissible to use a cast iron heat exchanger in a boiler for domestic hot water when

- **a** a direct storage cylinder is used
- **b** a descaling unit is installed
- **c** an indirect storage cylinder is used
- **d** the water is chemically treated.

8. The term dead-leg refers to

- **a** a painful right leg caused by kneeling
- **b** a pipe which is not part of a circuit
- **c** a pipe which is part of a circuit
- **d** a pipe for venting the system.

9. The lengths of dead-leg pipes are recommended in the British Standard

- **a** BS 2871
- **b** BS 7671
- **c** BS 6644
- **d** BS 6700.

10. The limitation of dead-leg length is designed to

- **a** allow smaller pipes and higher pressures
- **b** prevent overheating and reduce pressure
- **c** make the installation easy and cheaper
- **d** minimize wastage of water and maximize energy consumption.

11. A hot water storage cylinder in which the stored water returns to the boiler for reheating is referred to as

- **a** a combination cylinder
- **b** an indirect storage cylinder
- **c** a direct storage cylinder
- **d** a storage cylinder.

12. If a cast iron boiler was used with a direct storage cylinder it is likely that the water would become

- **a** discoloured
- **b** overheated
- **c** softened
- **d** hardened.

13. The most commonly used material used to manufacture pipes for hot water systems are

- **a** low-carbon steel and copper
- **b** stainless steel and plastic
- **c** copper and plastic
- **d** low-carbon steel and lead.

14. The material and type of tubing used in traditional installations is

- **a** copper tube manufactured to BS 2871 Table X
- **b** copper tube manufactured to BS 2871 Table Y
- **c** copper tube manufactured to BS 2871 Table Z
- **d** copper tube manufactured to BS 1363 Table 1.

15. A device connected to the cold feed pipe of an unvented hot water storage system to accommodate the increase in water volume when the stored water is heated up is called

- **a** an expansion tank
- **b** an expansion vessel
- **c** a relief valve
- **d** a tun-dish.

16. A non-return valve is installed in the cold feed pipe to an unvented cylinder after the branch leading to the cold water taps to prevent

a hot water flowing out of the cold water taps
b air from entering the system
c the expanding hot water returning to the mains
d the water from expanding on being heated.

17. The function of the pressure reducing valve is to

a prevent the water boiling away
b prevent water hammer in the pipework
c protect the expansion vessel
d give a constant downstream water pressure in the system.

18. The water expansion volume of the heated water flows into the expansion vessel and

a blows the air out of the vessel
b cools down and decreases its volume
c compresses the water into a smaller space
d compresses the air space by deflecting the bladder.

19. A device which will only permit water to flow in one direction is referred to as a

a non-return valve
b a pressure-relief valve
c a gate valve
d a service stop valve.

20. A device which is installed on a cold water service pipe to prevent a high mains pressure from causing damage to the system is referred to as

a a non-return valve
b a pressure-relief valve
c a pressure-reducing valve
d a globe valve.

21. If a situation arises in an unvented hot water storage system whereby the water pressure increases excessively, the device which prevents over pressurization is known as a

a a non-return valve
b a globe valve
c a pressure-reducing valve
d a pressure-relief valve.

22. The pressure-relief valve remains closed to normal pressure by means of

a a thermally sensitive phial
b a spring-loaded valve
c a dead-weight valve
d a screw-down valve.

23. If the water in an unvented storage cylinder reaches a temperature of 95°C a device known as a __________ opens and allows the water to flow to waste.

a thermal-relief valve
b pressure-relief valve
c non-return valve
d pressure-reducing valve.

24. The thermal-relief valve is opened by the expansion of a

a a large volume of water
b a bimetal rod
c heat-sensitive fluid in a phial
d spring-loaded valve.

25. Storage cylinders for use on unvented hot water storage systems are made from __________ than those cylinders used on vented systems.

a thinner sheet copper
b thicker sheet copper
c a different type of metal
d more expensive metals.

26. Unvented storage cylinders are supplied as a complete unit containing all the

a usual connections for pipes
b mandatory safety devices
c necessary BS numbers
d fixing instructions.

27. The function of the primary flow pipe is to convey hot water from the boiler to the

a open vent pipe
b rising main
c feed and expansion pipe
d storage cylinder.

28. The function of the primary return pipe is to convey cooled water from the storage cylinder to the

a tun-dish
b boiler
c expansion vessel
d rising main.

29. The primary flow and return pipes connect the

a boiler to the cylinder
b boiler to the cold water storage cistern
c cylinder to the cold water storage cistern
d cylinder to the draw-off points.

30. The primary circulation pipes are the means of conveying heat to the

a expansion cistern
b bathroom radiator
c stored hot water
d expansion.

31. The component that is an integral part of an unvented hot water system that is designed to prevent siphonic tendencies is referred to as

a a tun-dish
b an air gap
c a pressure relief valve
d a stop valve.

32. When it is necessary to provide a secondary return pipe to a hot water supply system the component used to circulate the water is a

a dead-leg
b water wheel
c by-pass
d pump.

33. The pipe on a domestic hot water storage system which is connected to the top of the cylinder is known as the

a secondary return pipe
b rising main
c secondary flow pipe
d primary flow pipe.

34. A secondary return pipe is added to a hot water supply system when the ______________ length exceeds those quoted in BS 6700.

a dead-leg
b primary flow pipe
c cold feed and expansion pipe
d primary flow pipe.

35. The 'point of delivery' is a term in the *Water Bylaws* which refers to the point where the secondary flow pipe connects to the

a lowest connection on the storage cylinder
b hot water outlet of the storage cylinder
c secondary return pipe
d feed and expansion cistern.

36. The length of pipe between the 'point of delivery' and the 'point of discharge' is referred to as a

a dead-leg
b secondary flow pipe
c secondary return pipe
d primary return pipe.

37. The advantage of installing an indirect storage cylinder is that the boiler heat exchanger can be manufactured from

- **a** aluminium or copper
- **b** glass or copper
- **c** cast iron or steel
- **d** stainless steel.

38. The indirect storage cylinder used on domestic hot water systems separates the primary heating water from the stored hot water by means of a

- **a** neoprene diaphragm
- **b** coil or annulus cylinder
- **c** separator valve
- **d** labyrinth valve.

39. When installing an indirect hot water storage cylinder it is necessary to provide a separate __________ for the primary circuit.

- **a** feed and expansion cistern
- **b** pressure relief valve
- **c** expansion valve
- **d** isolation valve.

40. One of the functions of an open vent pipe is to prevent siphonic tendencies developed within the system from

- **a** cooling the stored hot water
- **b** causing water hammer
- **c** reducing water pressure
- **d** collapsing the cylinder.

41. The 'point of discharge' on a hot water system is usually taken to be

- **a** the taps
- **b** the top of the cylinder
- **c** the ball valve
- **d** the drain-off valve.

42. The secondary return pipes are normally __________ those of the secondary flow pipes.

- **a** similar to
- **b** larger than
- **c** smaller than
- **d** equal to.

43. By installing a secondary return pipe to the system, the net effect is that the __________ is moved closer to the point of discharge.

- **a** storage cylinder
- **b** feed and expansion cistern
- **c** point of delivery
- **d** circulator.

44. The pump on the secondary flow pipe should be installed on a __________ to prevent the pump from impairing the flow of water to the taps.

- **a** plinth with anti-vibration mountings
- **b** opposite side of the cylinder cupboard
- **c** larger-diameter pipe
- **d** bypass checked by a non-return valve.

45. The open-vent pipe should be terminated directly over the __________ to collect any water 'spouted' from it.

- **a** guttering of the property
- **b** cold water storage cistern
- **c** bath
- **d** feed and expansion tank.

Now check your answers from the grid.

Q 1; d	Q 10; d	Q 19; a	Q 28; b	Q 37; c
Q 2; a	Q 11; c	Q 20; c	Q 29; a	Q 38; b
Q 3; c	Q 12; a	Q 21; d	Q 30; c	Q 39; a
Q 4; c	Q 13; c	Q 22; b	Q 31; a	Q 40; d
Q 5; d	Q 14; a	Q 23; a	Q 32; d	Q 41; a
Q 6; a	Q 15; b	Q 24; c	Q 33; c	Q 42; c
Q 7; c	Q 16; c	Q 25; b	Q 34; a	Q 43; c
Q 8; b	Q 17; d	Q 26; b	Q 35; b	Q 44; d
Q 9; d	Q 18; d	Q 27; d	Q 36; a	Q 45; b

10

DOMESTIC HEATING SYSTEMS

To tackle this chapter you will need to know:

- the domestic heating system – combination boiler;
- the domestic heating system – sealed system;
- the domestic heating system – fully pumped 'S' Plan;
- the domestic heating system – fully pumped 'Y' Plan;
- the various components found on domestic heating systems.

GLOSSARY OF TERMS

Automatic air vent (AAV) – an automatic device which is at high points on systems where air cannot naturally vent out from the system. One type in common use operates by means of a needle valve attached to a float contained within a small chamber. When air collects in the chamber the float falls with the water level thus pulling the needle valve from its seating and therefore releasing the air. As the water expels the air the water level and the float rise and reseal the valve.
Close coupled C/F and O/V – a method of connecting the cold feed and expansion pipe and the open vent pipe to the same point on the system. The position is on the suction side of the pump within 150 mm. The suction pressure of a centrifugal pump is very low compared to the delivery pressure. Therefore this position does not create 'spouting' of the open vent. It also means that the only negative pressure in the system is the length of pipe between the cold feed and expansion pipe connection and the pump.
Cold feed and expansion cistern – a component which has several functions within the system, primarily to fill the system, to top up the system in the event of leaks and to accommodate the expanding water on heating. It is fed with water through a float-operated valve.
Cold feed and expansion pipe – a pipe used to fill the system and also to convey the expanded water volume out of the system into the feed and expansion cistern. In a sense it is a pressure relief valve and therefore should have valves placed in it to prevent the system becoming sealed. The point where it connects to the system is taken as the 'neutral point'; i.e, the point at which the water pressure is neither positive nor negative and there is effectively no water movement.
Column radiators – column radiators are available manufactured from die-cast aluminum. They tend to be decorative in appearance and are available in many different coloured finishes. Their design is based on the old cast iron column and hospital radiators. Although referred to as radiators the majority of heat given off is by means of convection.
Combination boiler – an appliance that incorporates an instantaneous hot water heater and a system boiler, hence the name combination boiler. Combination boilers have become very popular, particularly in the case of 'retrofits' into existing systems. Either oil or gas fuels these boilers. The operation of the boiler is by a series of heat exchanges and differential pressure switching devices. The plate heat exchanger is now the preferred choice. It is possible to operate the hot water side without the central heating on or in fact connected.
Common flow – on fully pumped systems the flow of hot water from the boiler to all the various zones of the system is carried in one pipe to the pump and thereafter it splits to feed the zones of the system. The C/F and O/V pipes connect to the common flow pipe prior to it entering the suction side of the pump.
Common return – a pipe that carries the total mass flow rate of the system. The primary return pipe and the heating circuit return pipe both join into this common return pipe. It is important that all the heating returns are connected together before the primary return pipe is connected. If this is not the case then it is probable that short circuiting will occur through the radiators when hot water only is selected on the programmer.
Convector radiator – modern panel radiators are manufactured from two sheets of pressed steel which are welded together around the seams. The pressings form a top and bottom manifold with waterways flowing between them. Each radiator has a stated heat output which is equated to a surface area. The convector radiator produces a larger surface area by adding a corrugated sheet of steel spot welded to the back panel. Therefore size for size its output is much greater than the plain panel radiators. These types are available as single or double panel.
Cylinder thermostat – a thermostat that controls the stored hot water temperature, and enables the water temperature to be set independently from the heating water temperature. It is an integral part of the fully pumped system controls. It will control either a 'two-part' motorized valve or determine where the valve in the three-part mid-positional motorized valve will rest.

Double check valve – sealed heating systems, including those with a combination boiler installed, are filled and topped-up by means of a 'filling loop'. The *Water Bylaws* state that a double check valve is placed between the system pipework and the filling loop. Its function is to prevent the system expansion volume expanding back into the cold mains. It also prevents the system pressure deflating.

Expansion vessel – a component associated with sealed systems, i.e. systems not open to atmospheric pressure. The vessel is divided into two spaces separated by a flexible membrane; one space is connected to the waterway of the system and the other is filled with air at a predetermined pressure. As the system expands the increase in volume deflects the membrane and thus compresses the air in the air space. Therefore the water expansion volume occupies the majority of the air space. The capacity of the vessel is sized on the heat output of the system.

Fan convector – specially designed units for domestic heating systems. A popular model is made specifically to be installed under kitchen units – the 'kick space'. These units are thermostatically controlled either integrally or by remote sensor. Some fan convectors have low limit thermostats to prevent the fan operating until the flow water reaches a high temperature.

Filling loop – a flexible, reinforced hose-type connection permissible by the *Water Bylaws* which connects the rising main to the heating pipework. It is used to fill and top up the system. It should be removed after the system has been filled.

Heating flow – a pipe that supplies fresh heated water from the boiler to the radiator circuits.

Heating return – a pipe that conveys the cooled water exiting the radiators back to the boiler for reheating.

Heating zone – by utilizing motorized valves and room thermostats it is possible to split heating systems up into individual zones for greater control and economy.

Open vent pipe – a pipe that is connected to the primary or secondary flow pipe at its highest point, from where it is taken to a point where it can discharge any water 'spouted' from the system over the feed cistern. This pipe is open to atmospheric pressure and therefore prevents the water boiling point from going higher than 100°C. It also permits air to escape automatically from the system. If connected to the suction side of the pump it is possible for air to be drawn into the system

Panel radiator – a radiator constructed like the convector radiator, but without the additional corrugation on the back panel. All types of radiators have four connections giving several possibilities for connecting to the system. Most common in domestic heating is 'bottom opposite ends'. One of the top connections will be an air release valve.

Pressure-relief valve (PRV) – an essential safety device used to maintain the pressure of the system constant. The pressure of the system acts upon a valve, which is spring-loaded to a predetermined pressure value. Should the water pressure increase above the relief pressure setting, the valve will lift and water will flow to waste until the status quo is regained whereupon the valve will reseat. This process limits the amount of water that flows to waste. The exhaust valve from the valve must be terminated to outside air.

Programmable room thermostat – an electronic programmer and a room thermostat. It is often used with combination boiler installations to control the operating times of the heating and also the air temperature of the room in which it is installed. It will not affect the domestic hot water supply.

Pump (circulator) – an electrically driven machine used to circulate water around the system. It generates a circulating pressure or head capable of overcoming the flow resistances of the total system. Usually of the centrifugal type, modern pumps have variable head settings for great flexibility.

Room thermostat – an automatic switching device, which senses the ambient air temperature around it. It is used to switch a pump on and off or to operate the movement of a motorized valve. Modern room thermostats incorporate an 'anticipator' which is a small resistor, which emits heat directly under the air heat sensor. This has the effect of operating the device at 2°C less than the set point temperature of the thermostats. The net effect is a constant air temperature within the room.

Sealed heating system – a system which is closed to atmospheric pressure; no open vent or feed and expansion cistern is installed. Instead an expansion vessel is connected to the return pipe close to the boiler. A pressure relief valve is also installed to release excess pressure when it prevails. The system is filled by means of a filling loop. The boiler and radiator valves should be capable of withstanding high pressure. These systems normally operate around 2 bar gauge pressure when hot.

'S' plan – fully pumped – a type of heating system developed for the domestic market which is fully pumped throughout and controlled automatically. It utilizes two two-part motorized valves, one to control the temperature of stored hot water and the other to control the air temperature in the home. A thermostat controls each valve. The motor-head contains micro-switches, which in turn control the feed to the boiler controls and the pump.

System boiler – a straightforward heat generator with a combustion chamber and heat exchanger used for both heating and the hot water requirements. Fueled by any of the common energy sources including electricity, there are suitable models for both sealed and open systems. Installation is controlled by current Regulations.

Thermostatic radiator valves – a component which is essentially a radiator valve, which is usually installed on the flow pipe into the radiator. The valve gearhead is replaced by a thermally sensitive head which expands and contracts due to the heating and cooling of the air around it. This movement controls the flow of water to the radiator. These valves provide individual control to each radiator and they can greatly improve economy of use.

Three-port mid-positional valve – a valve similar to the two-port valve but more complicated. The valve has a common flow inlet and two zone outlets – one for hot water and one for heating. In the valve head are two micro-switches. Two thermostats control the valve. Because it can only close one port off at any one time, when both thermostats are satisfied the two micro-switches cut off the power supply to the boiler and pump. The valve will remain in the power-Off position until heat is called for. If, for instance, the cylinder thermostat calls for heat, the valve will move across and close off the heating port so preventing water flowing to the radiators. At this point power will be restored to the boiler and pump. The sequence will be the same if the opposite occurs.

Two-port motorized valve – an on/off device which will operate when a power supply is provided by a thermostat. The valve is spring-loaded and therefore returns to the closed position when the power supply is stopped. Micro-switches in the motor-head switch power to the boiler control and the pump when the valve is in the open position.

'Y' plan – fully pumped – an alternative type of heating system which is fully pumped but incorporates a three-port mid-positional valve as described above. The remainder of the system is the same as the 'S' plan.

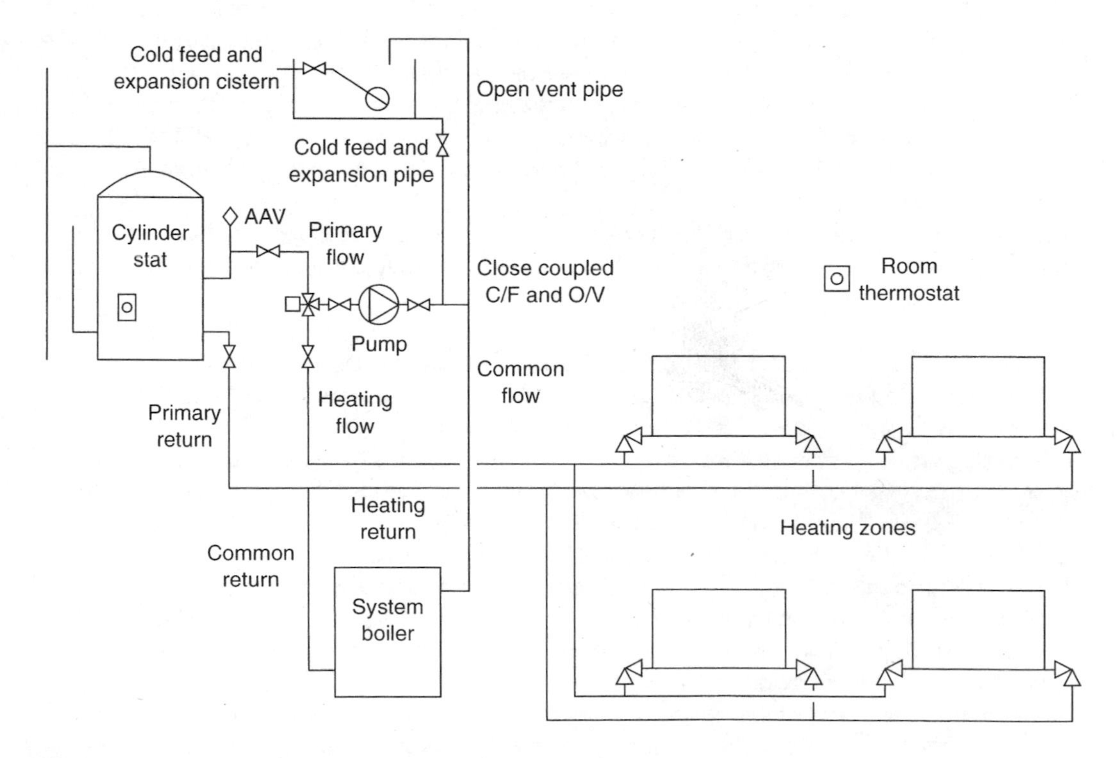

Fig. 10.1 'Y' plan – fully pumped using three-port mid-positional valve

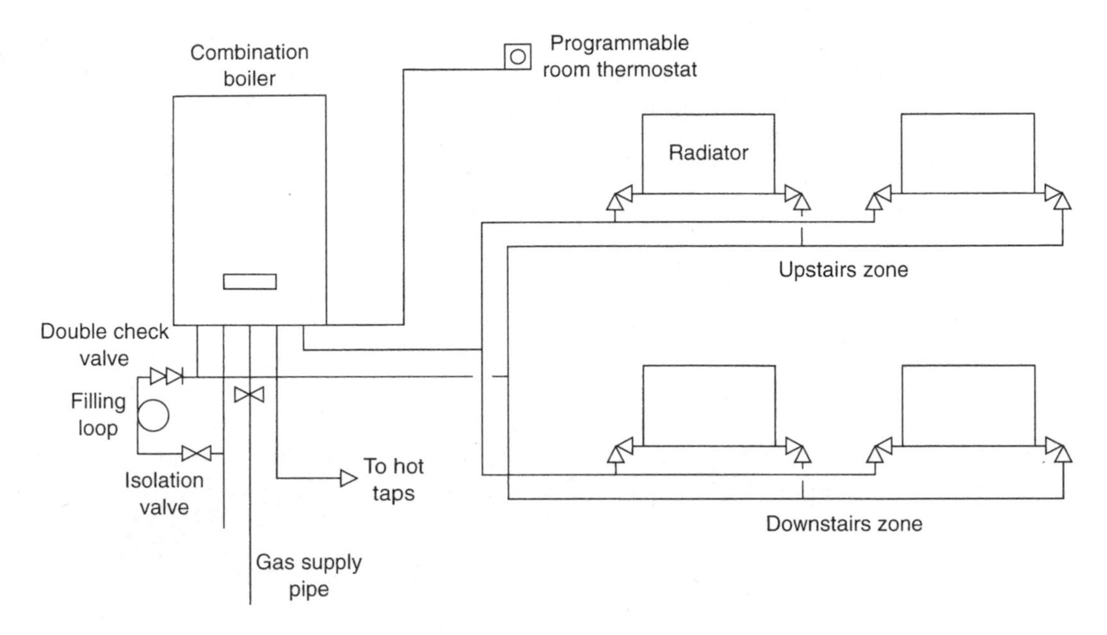

Fig. 10.2 Combination boiler installation

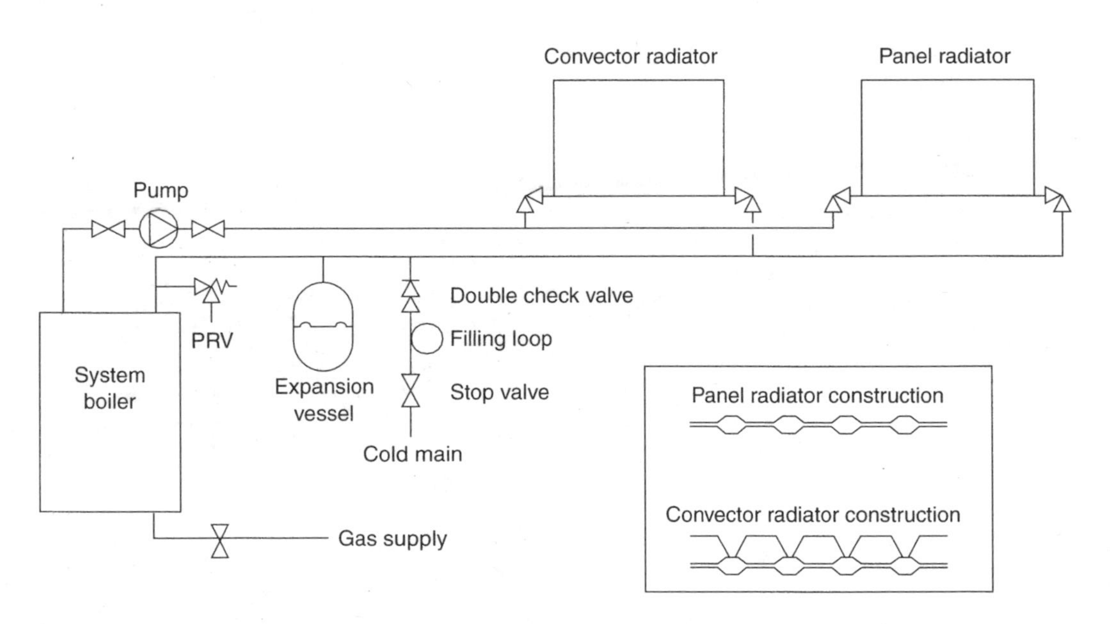

Fig. 10.3 Sealed heating system

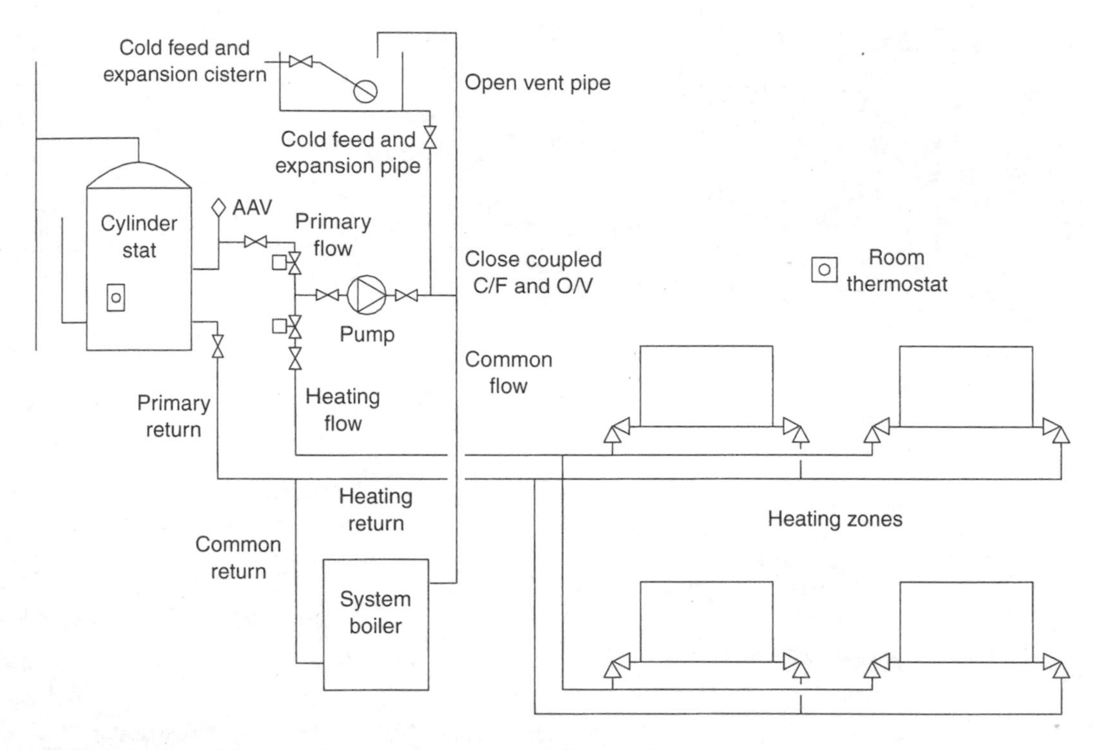

Fig. 10.4 'S' plan – fully pumped using two two-port motorized zone valves

Now try and answer these questions

1. A device which is designed to prevent the expanding water in a sealed system pushing back into the rising main is called
   - **a** a pressure-relief valve
   - **b** a service stop valve
   - **c** a double check valve
   - **d** a quick-acting valve.

2. The pipe on a heating system which conveys hot water from the boiler to the inlet to heat emitters is referred to as
   - **a** the open-vent pipe
   - **b** the cold feed pipe
   - **c** the heating flow
   - **d** the heating return.

3. A component which is designed to control the air temperature within a room by sending a signal to open or close a motorized valve is commonly referred to as
   - **a** a room thermostat
   - **b** a cylinder thermostat
   - **c** an off/on switch
   - **d** a time clock.

4. In fully pumped heating systems it is very important to connect all the heating returns before the primary return pipe is connected to the 'common return' to prevent
   - **a** over expansion of the water
   - **b** localized cooling
   - **c** airlocks in the radiators
   - **d** short circuiting through the radiators.

5. When the filling of a sealed heating system has been completed the connection between the rising main and the heating pipework, known as ________ , should be removed.

a a double check valve
b a filling loop
c an expansion vessel
d a feed and expansion vessel.

6. The open vent pipe will enable the system to maintain ______ and therefore prevent the water boiling point exceeding 100°C.

a atmospheric pressure
b a higher pressure
c subatmospheric pressure
d a lower pressure.

7. A type of heating system which has neither a feed and expansion cistern nor an open-vent pipe installed is referred to as

a an open system
b a gravity system
c a sealed system
d a dual system.

8. On sealed heating systems an important safety device which prevents excessive over pressurization is called

a an open-vent pipe
b an automatic air vent
c an expansion vessel
d a pressure relief valve.

9. A pump is a mechanical device which is sized to generate sufficient pressure to overcome the __________ of the system.

a expansion pressure
b total flow resistances
c static head
d hydraulic gradient.

10. A ______ is a device which opens or closes in response to signals from a thermostat.

a motorized valve
b pressure-relief valve
c automatic air vent
d filling loop.

11. A type of heating boiler which eliminates the need for a hot water storage cylinder is referred to as a

a system boiler
b combination boiler
c balanced flued boiler
d free-standing boiler.

12. One method used to give additional control and economic use of heating systems is to split the system up into separate

a heating systems
b one-pipe systems
c two-pipe systems
d heating zones.

13. An integral component of a fully pumped system's controls which maintains the stored hot water temperature is referred to as a

a room thermostat
b cylinder thermostat
c programmable room stat
d controller.

14. The opening and closing of a motorized valve controls the flow or power to both the pump and the boiler by means of ______ in the motorhead.

a micro-switches
b thermostats
c double-pole switches
d fused spurs.

15. The components installed in sealed heating systems should be capable of

- **a** withstanding low pressure
- **b** receiving chrome-plating
- **c** taking heavy knocks
- **d** withstanding high pressures

16. An electric control device which is both capable of controlling the room temperature and the on/off times of a system is known as

- **a** a programmable room thermostat
- **b** a room thermostat
- **c** a heating programmer
- **d** a time clock.

17. When installed on a combination boiler system a programmable room thermostat will not affect the operation of

- **a** the system pump
- **b** the gas rate of the boiler
- **c** the domestic hot water supply
- **d** the heat input.

18. A modern room thermostat has a device called ______ which makes it more efficient at controlling the room air temperature.

- **a** an anticipator
- **b** an infra-red sensor
- **c** a movement sensor
- **d** a plastic fixing bracket.

19. In the 'Y' plan heating system when both thermostats are satisfied, micro-switches in the motorhead will

- **a** switch off the power to both the thermostats
- **b** switch off the power to the boiler and pump
- **c** divert the flow water back to the boiler
- **d** keep only the domestic hot water heated.

20. The three-port motorized valve is capable of

- **a** increasing the flow rate of the system
- **b** decreasing the flow rate of the system
- **c** closing both of the two outlet ports
- **d** closing only one of the outlet ports.

21. The exhaust pipe attached to a pressure relief valve should terminate ______ for the purpose of safety.

- **a** to the outside air
- **b** directly beneath the boiler
- **c** over a convenient sink
- **d** directly over the feed and expansion cistern.

22. Heating zones are automatically controlled by

- **a** pressure relief valves and open vent pipes
- **b** time switches
- **c** motorized valves and thermostats
- **d** thermostatic radiator valves.

23. A device which can be fitted on to a radiator to provide individual control is referred to as

- **a** a thermostatic radiator valve
- **b** a two-port motorized valve
- **c** a three-port motorized valve
- **d** a lockshield valve.

24. When a centrifugal pump is operating it produces

- **a** a large amount of noise and vibration
- **b** a large amount of useful heat
- **c** a high delivery pressure/low suction pressure
- **d** a low delivery pressure/high suction pressure.

25. The concept of 'close coupled cold feed and open vent pipes' enables the two pipes to be connected to the

- **a** common flow pipe
- **b** common return pipe
- **c** heating flow
- **d** same point on the system.

26. An alternative to the panel radiator, heat emitters are available manufactured from die-cast aluminum, referred to as

- **a** double heat exchangers
- **b** column radiators
- **c** coils
- **d** ray strips.

27. The most common method of piping radiators on domestic heating systems is the

- **a** bottom opposite ends method
- **b** top and bottom opposite ends method
- **c** top and bottom same ends method
- **d** top opposite ends method.

28. On small fan convector units a component attached to the flow pipe known as ________________ prevents the fan running until the water reaches a high temperature.

- **a** a gate valve
- **b** a two-port valve
- **c** a low-limit thermostat
- **d** a room thermostat.

29. Convector radiators have an additional corrugated sheet welded to the back panel to

- **a** decrease the total surface area
- **b** increase the total surface area
- **c** make them look modern
- **d** to prevent dust gathering.

30. The 'S' Plan heating system is a fully pumped system comprising two circuits controlled by

- **a** three-port motorized valves
- **b** thermostatic radiator valves
- **c** lockshield valves
- **d** two-port motorized valves.

31. A device for venting air from pipework without using a key is called

- **a** a screw-down valve
- **b** a double check valve
- **c** an automatic air vent
- **d** a drain-off valve.

32. A component part of an open system which has several functions, primarily to fill the system, is referred to as

- **a** an automatic flushing cistern
- **b** a cold feed and expansion cistern
- **c** a filling loop
- **d** an expansion vessel.

33. The only type of valve that is recommended to be placed in the feed and expansion pipe of an open system is

- **a** no valve at all
- **b** a gate valve
- **c** a ball valve
- **d** a motorized valve.

34. The function of an expansion vessel installed on a sealed heating system is to accommodate the

- **a** volume of water expanded on heating
- **b** likely extensions to the system
- **c** increase in pressure on heating
- **d** decrease in pressure on heating.

35. The point at which the cold feed and expansion pipe connects to the system pipework is the ________ where the pump pressure is neither positive or negative.

a coldest point
b hottest point
c drain-off point
d neutral point.

36. It is common in fully pumped heating systems to find the cold feed and open vent pipes connected into the

a rising main pipe
b common flow pipe
c common return pipe
d exhaust pipe.

37. Modern panel radiators are manufactured from ________ which are welded together.

a one sheet of pressed steel
b die-cast alloy sections
c two sheets of pressed steel
d cast iron sections.

38. Under certain circumstances if the open vent pipe is connected to the suction side of the pump ________ can be drawn into system.

a cold water
b products of combustion
c air
d water vapour.

39. By installing thermostatic radiator valves in a system, there is likely to be a reduction in the

a running costs
b heat output
c efficiency of the system
d pump noise.

40. The operation of a combination boiler is controlled by a series of

a timeclocks and thermostats
b manual valves and switches
c heat exchangers and timeclocks
d heat exchangers and differential switches.

41. Several modern combi-boilers utilize

a coil heat exchangers
b plate heat exchangers
c annulus heat exchangers
d cross-over heat exchangers.

42. The motorized two-port valve returns to the closed position when the call for heat is satisfied by means of

a a shunt pump
b a reverse acting motor
c an auxiliary motor
d a return spring.

43. The motorized three-port mid-positional valve is controlled by two

a thermostatic radiator valves
b thermostats
c programmers
d pipeline heat sensors.

44. Fully pumped heating systems enable greater flexibility by having independently controlled

a zonal circuits
b boilers
c pumps
d pressure vessels.

45. A system boiler cannot provide ________________, but it will heat a storage cylinder for hot water supply.

a tea making facilities
b water for intermittent use
c instantaneous hot water
d boiling water.

Now check your answers from the grid.

Q 1; c	Q 10; a	Q 19; b	Q 28; c	Q 37; c
Q 2; c	Q 11; b	Q 20; d	Q 29; b	Q 38; c
Q 3; a	Q 12; d	Q 21; a	Q 30; d	Q 39; a
Q 4; d	Q 13; b	Q 22; c	Q 31; c	Q 40; d
Q 5; b	Q 14; a	Q 23; a	Q 32; b	Q 41; b
Q 6; a	Q 15; d	Q 24; c	Q 33; a	Q 42; d
Q 7; c	Q 16; a	Q 25; d	Q 34; a	Q 43; b
Q 8; d	Q 17; c	Q 26; b	Q 35; d	Q 44; a
Q 9; b	Q 18; a	Q 27; a	Q 36; b	Q 45; c

11

BELOW-GROUND DRAINAGE

To tackle this chapter you will need to know:

- the types of systems used in below-ground drainage;
- the means of access to drainage systems;
- the materials used in the production of drainpipes and fittings;
- the definition of drainage systems;
- the method used to test drains.

GLOSSARY OF TERMS

Access – access to drains can be through one of the four means listed below

i rodding eye
ii access fitting
iii inspection chamber
iv manhole

and should provide a means for the clearance of blockages. Suggested locations are at a point near to the head of a drain, at a change of direction or gradient, where a change in diameter occurs, or at a junction of drains or branch drains.

Air Bag Stoppers – alternatives to mechanical expanding drain plugs. They are constructed from rubberized canvas with a rubber hose attached to one end for the purpose of admitting air from a hand pump. When fully inflated within a pipe they completely seal off the interior of the pipe for testing purposes.

Anti-flooding valve – an inline trap-type device, which has a copper ball-float in the upstream side of the trap. At the inlet to the valve is situated a rubber seating. If a backflow is present from the sewer side of the trap the copper float is pushed on to the seating which prevents the flow of liquid in both directions. The float will only be released when the flooding subsides.

Backdrop manholes – where the inclination of the ground is greater than the gradient of the drain/sewer pipe the use of backdrop manholes enables the drain's gradient to be maintained. In the backdrop manhole the upstream pipe enters at the top of the chamber and is then piped to the invert off the downstream pipe exiting the manhole. The drop pipe can either be outside the chamber or within it, depending on the material of the pipe, e.g. clay pipe would be laid outside the chamber, or alternatively cast iron pipe would be more suited for installation inside the chamber.

Back (vertical) inlet gully – a gully with a dedicated connection for either a waste or rainwater pipe without it having to discharge over a gully grate. This has the effect of preventing blockages due to an accumulation of leaves or debris on the grate.

Bedding – the preparation to the base of the trench prior to the drainpipe being laid. The form and depth of the bedding is dependent on the material of the pipe and the type of soil and relationship to a building.

Bedding classification

Classification	Pipe	Bedding	Backfill
Class A	Rigid pipes Vitrified clay	100 mm min concrete and part way up the pipe barrel.	Selected soil cover for 300 mm Main backfill above.
Class B	Rigid pipes Flexible pipes	100 mm min granular material (completely cover UPVC pipe) (half diameter clay and cast iron pipes).	Selected soil in 100 mm layers.
Class D	Rigid pipes Flexible pipes	No bedding; lay directly onto trench base.	300 mm selected soil cover Main backfill above.

British Standards for cast iron drainware

BS number	Definition
BS 78: Part 1, 1961	Spigot and socket pipes. All-purpose pipe, 75 to 300 mm diameters available in lengths of 2.74 m to 3.66 m.
BS 437: Part 1, 1970	Spigot and socket pipe for both foul and surface drains, 48 to 225 mm diameters available in lengths 1.8, 2.74 and 3.66 m.
BS 1211: 1981	Centrifugal spun iron pressure pipes. All-purpose pipe, 75 to 900 mm diameters available in lengths of 3.66, 4.0, 4.88 and 5.5 metres.
BS 4622: 1975	Cast iron pipe and fittings for drainage systems, 80 to 700 mm diameters available in a selection of lengths.

British Standards for clay drainware

BS number	Definition
BS 65 and BS 540: 1971	Drain and sewer pipes including surface water.
BS 539: 1968	Fittings for use with clay drainpipes.
BS 1143: 1955	Salt glazed with chemical-resistant properties.
BS 1196: 1971	Land drainpipe.

British Standard 8301: 1985 – the British Standard covering below-ground drainage and should be referred to when designing systems.

Building Regulations – Approved Document H of the 1985 Building Regulations, for England and Part M Building Standards, for Scotland prescribe the requirements of drainage systems in terms of: access, gradient, design considerations, bedding and proximity to buildings. These documents should be referred to when working on drainage systems.

Cast iron – cast iron is a very durable material, which can be formed into both pipes and fittings. Pipe is either spun cast iron to BS 1211 or sand-cast to BS 437; Part 1, 1970. All fittings are produced using the sand-cast method. Fittings are produced in a very comprehensive range from simple obtuse bends to complicated inspection chambers. Cast iron drainage systems are traditionally laid beneath buildings because of their strength.

Caulked joint – a traditional plumbing method of making a watertight joint on cast iron pipe systems including drainage. The space between the spigot end of a pipe or fitting and the socket end of a pipe or fitting is filled with lead. After the spigot end has been inserted into the socket a round of 'gaskin' (soft jute rope) is rammed home to the back of the socket. This has the effect of bringing the two components into line. Molten lead is then poured into the joint. After the lead has cooled, special tools are used (caulking irons) to ram the lead into the joint, and in so doing make the joint watertight. This type of joint is a 'rigid joint' and therefore where movement is likely a 'flexible' joint should be provided – see victaulic joint.

Combined system – a system of below-ground drainage where both surface water and foul water are removed through one system of pipework. At times of high rainfall the system is flushed and therefore cleaned out at regular periods. With this system it is not possible to connect a soil or waste discharge to a surface water drain.

Drain plugs – expanding devices, which increase in diameter when compressed by the turning down of a brass wing nut. The rubber seal expands to fill the internal bore of the pipe. The plugs are used in the testing process for both above-ground and below-ground drainage systems. The screw thread, which carries the wing nut, is hollow, permitting connection to the test equipment.

Gradient – the rate of fall of a drain necessary to permit the drain to have a flow of sewage whereby all solids are cleared from the drain on flushing. The drain must flow at a velocity that is sufficient to keep it free from blockage – the 'self-cleansing velocity'.

Grease trap – a specially designed gully for large kitchens or food process plants, etc. It traps grease and other undesirables from entering the drain by catching them in a removable tray made from mesh or drilled metal sheet. The tray is removed at regular periods and the contents disposed of elsewhere.
Gully – a drainpipe fitting with a water seal used to connect waste discharge pipes and rainwater pipes to a drainage system. Soil discharge pipes must never be connected to a drainpipe via a gully of any description; this will prevent blockages occurring. The water seal prevents the egress of foul air emanating from the drain. Drainage systems are ventilated through discharge stacks (Chapter 12). There is also a trapless gully available for use on surface water drains only.
High-strength clay pipe – a manufacturer (Hepworths) has developed a process of manufacturing clay pipe which is capable of providing a reduction in length-to-weight ratio, and greater strength. There is also a higher degree of diameter accuracy and consistency. The system is referred to a 'SuperSleve', which has a nominal bore of 100 mm diameter. There is also a wide range of fittings. Jointing is made using couplings manufactured from polypropylene with synthetic rubber 'O' rings; the joint provides flexibility.
Incline (I) – a slope, gradient or fall. This is usually expressed as

$$\frac{H}{L} = \frac{\text{head in metres}}{\text{length in metres}} = \text{incline}$$

Maguire's rule for an incline is: 1 (one) in the distance L

e.g. firstly; take the diameter of the drain to be laid
secondly; divide this by the constant 2.5

For a 100 mm drainpipe we get

$$\frac{\text{100 mm diameter drain}}{\text{2.5 (constant)}} = 1{:}40$$

This incline (1:40) means that every 40 m length of drain will fall over a distance of 1 m. This will produce a 'self-cleansing velocity' for all 100 mm drains.
The self-cleansing velocity is usually accepted as 1 m/s.

Therefore for a 150 mm drain we find:

$$\frac{\text{150 mm diameter drain}}{\text{2.5 (constant)}} = 1{:}60$$

From these ratios simply dividing 40 into 1 can derive a multiplying factor. This will give a factor of 0.025 for 100 mm drainpipes and 1/60 or 0.0167 for a 150 mm drainpipe.

Example 11.1 A 100 mm drainpipe is to be laid over a 10 m distance. Calculate the total amount of fall.

$0.025 \times 10 = 0.25$ m or 250 mm

Example 11.2 A 150 mm drainpipe connecting a discharge stack to a manhole is 20 m in length. Calculate the total fall of the drainpipe.

$1 / 60 = 0.0167$
$0.167 \times 20 = 0.334$ m or 334 mm

Inspection chamber – an access chamber up to 1 m in depth with access covers. It is not possible to work in an inspection chamber, work is carried out from the surface.

Invert – the internal base of the drain pipe. The invert of the drain is used to set up the levels to achieve the necessary gradients.

Maguire's rule – a simple equation used to calculate the amount of fall in the drain for a given length and is dependent on diameter. This in effect will produce a 'self-cleansing velocity'.

Manhole – a structure which provide a means of access to drains for the purpose of cleaning and inspection. Construction is varied, ranging from brick-built construction and pre-cast concrete ring sections, to small plastic mouldings. Manholes are situated along the length of the drain/sewer at distances of not more than 90 m, at changes in direction, where branch drains connect to a main drain or where there is a significant change in gradient. Manholes are access points, which are more than 1 m in depth and provide a means for cleaning and inspecting drains below ground level.

Partially separate system – a system of below-ground drainage, which is basically a separate system, used in certain circumstances, where for instance, it would be inconvenient or costly to lay a drainpipe specifically to pick up an isolated rainwater pipe. It is possible under these conditions to connect the rainwater drop to the foul water drain. Under no circumstances is it permissible to connect a soil or waste system to a surface water drain.

Petrol interceptor – a three-chambered trap which is used to connect a surface water drainage system in a large car repair workshop or car park area and a public sewer to prevent oil and fuel residues entering the sewage system. Water moves through the chambers by means of a series of dip-pipes.

Private drain – a privately owned drain serving only the property it is laid in.

Private sewer – a common sewer laid through the grounds of individual properties and shared as a means of conveying sewage to the public sewer. The responsibility for these sewers is shared equally amongst the properties connected to them. These sewers could be any of the three systems described dependent upon the local conditions.

Public sewer – the main drain/sewer, which is predominately laid beneath roads and streets. Public sewers are the responsibility of the local authority or the water undertaking. A charge for their use and maintenance is part of the water bill imposed on a property.

Puddle flange – either a clamp-on or cast-on flange on a section of cast iron drainpipe which is to be cast into a concrete structure below ground level. It acts as a water barrier preventing ground water entering the building.

Push-fit joint – a joint which is used in various forms on cast iron, UPVC and clay pipes, it employs the rubber 'O' ring method of jointing. It is the 'O' ring that provides the seal. This type of joint provides flexibility and protects the pipework from being sheared when land or building movement takes place. These joints will enable a pipe to deflect up to 5° generally, thereby protecting the water tightness of the joint.

Rainwater shoe – a fitting with no water seal that can only be used where a rainwater drop-pipe is connected to a surface water drain.

Rodding eye – a means of access to a drainpipe without constructing a manhole/inspection chamber. It is usually to be found at the head of a length of drain with only one branch drain connected. The drainpipe is brought to ground level by means of a long-radius bend and a length of pipe where a sealed access cover is provided.

Salt glazed clay – the traditional material for small-diameter drainage systems, which are laid in open ground. Both pipe and fittings are produced in various types and lengths to enable installation in all configurations. Joining is effected either by placing a round of gaskin and then pointing off with mortar or, of late, using push-fit collars with 'O' rings.

Separate system – a system of below-ground drainage where the surface water and the foul water are conveyed in totally separate pipework systems. This has the advantage of reducing the amount of sewerage treatment and therefore cost. The surface water is normally discharged into a natural watercourse, e.g. stream or river.

Testing – Air testing: to test drainage systems using air the part of the system being tested should be sealed off and the system subjected to a pressure equal to 100 mm water gauge. Five minutes for temperature stabilization should be allowed before carrying out a further five-minute test period. During the test period the water gauge should not fall more than 25 mm. This type of test is applied to foul drains. Water testing: to test drainage systems with water the system should be subjected to a 1.5 m head of water at the highest end. Allow to stand for 2 hours to enable absorption and top up as necessary. The actual test is 30 minutes and the loss of water rate is dependent on the diameter and length being tested. This type of test is applied to surface water drains.

Unplasticized polyvinylchloride (UPVC) – the modern material used to produce both pipes and a wide range of drainage fittings. UPVC pipes are manufactured to BS 4660: 1971, which covers pipes and fittings of 100 mm diameter. There are other diameters produced up to 200 mm. The pipe is coloured brown and is thicker walled than the type used for above-ground use. Jointing is by the push-fit 'O' ring method. These joints allow for movement to occur without causing damage. Long discharges of hot waste can cause distortion. UPVC pipes require a 100 mm granular material for bedding.

Victaulic joint – a type of flexible mechanical joint for cast iron drainpipes which pass through concrete walls of both manholes and structural walls. They permit the drain to flex when building settlement takes place without shearing the pipes. The joint is a cast iron sleeve with two end flanges and rubber compression rings. Four full-length galvanized bolts hold the component parts of the joint in place.

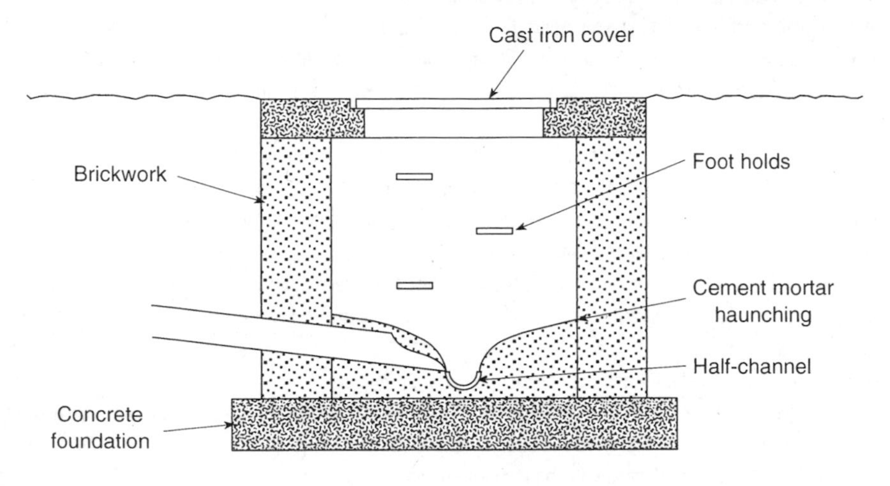

Fig. 11.1 Manhole

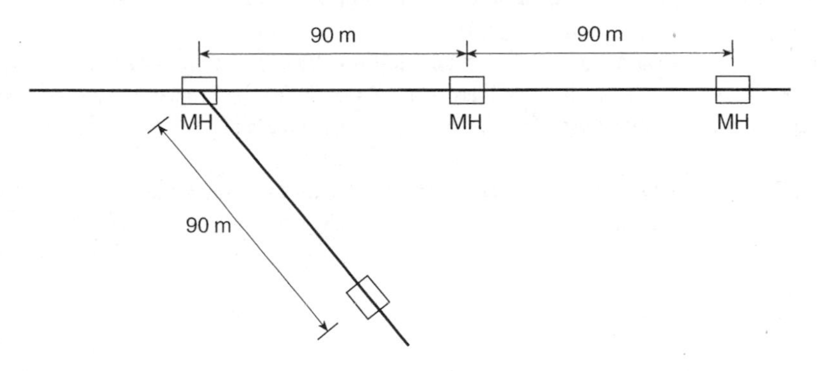

Fig. 11.2 Access to drains

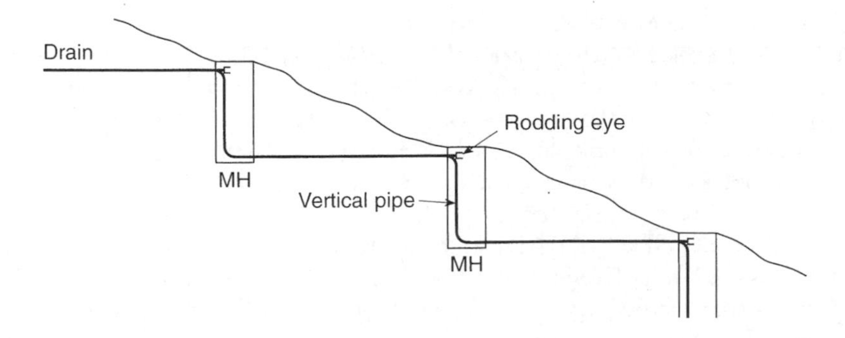

Fig. 11.3 Backdrop manholes

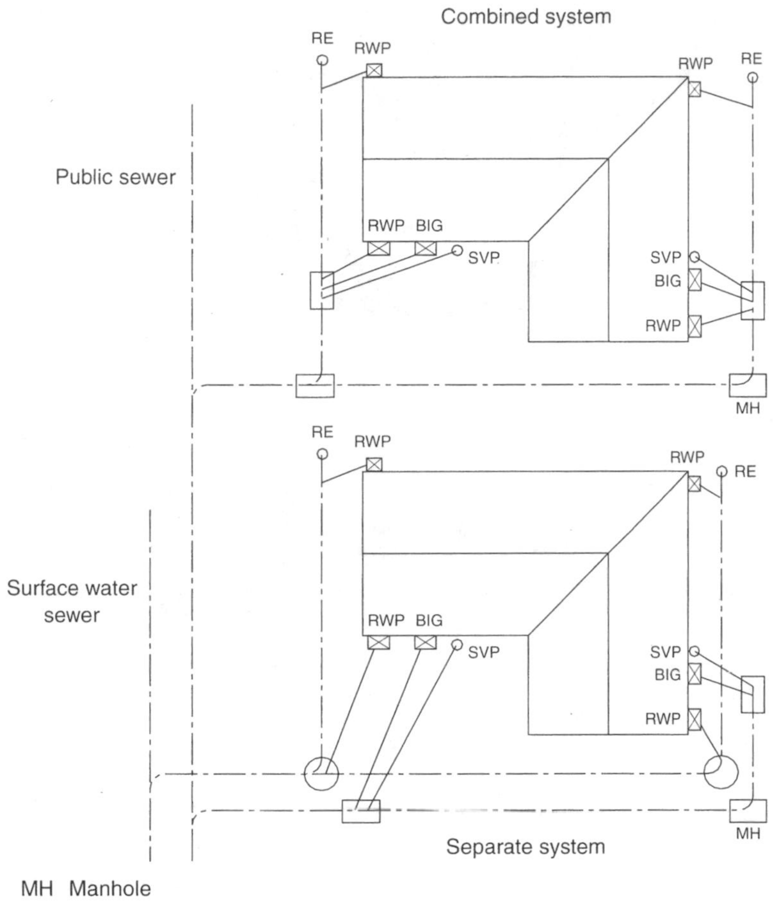

MH Manhole
RE Rodding eye
RWP Rainwater pipe
SVP Soil and vent pipe
BIG Back inlet gully

Fig. 11.4 Drainage systems

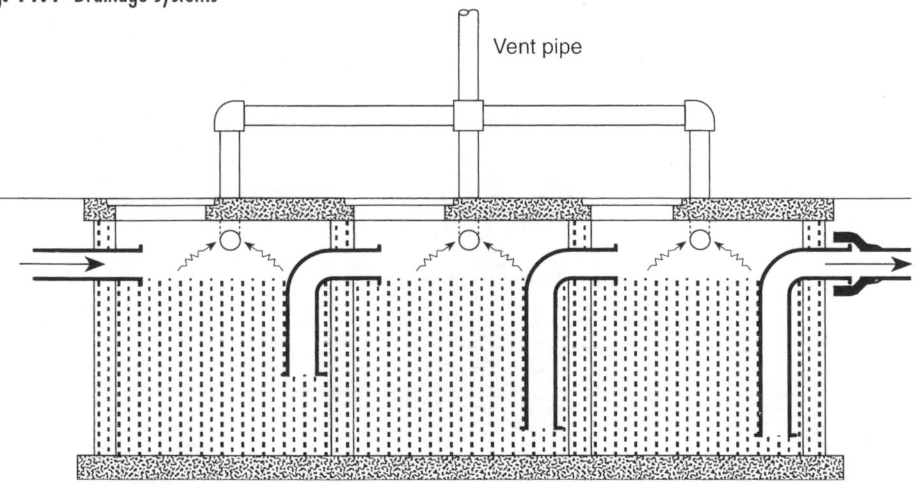

Fig. 11.5 Petrol interceptor

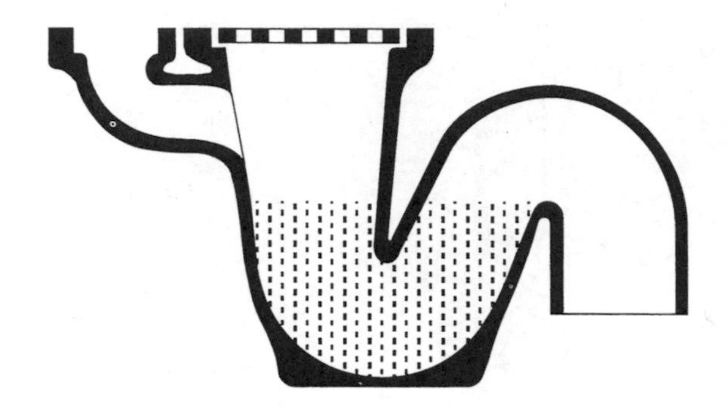

Fig. 11.6 Back (vertical) inlet gully

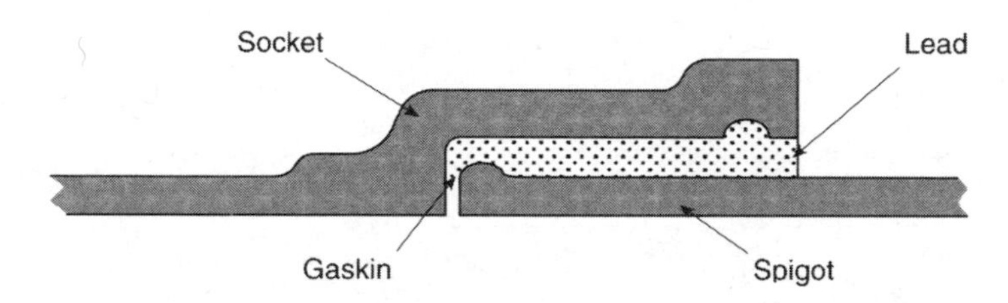

Fig. 11.7 Lead caulked joint on cast iron pipe

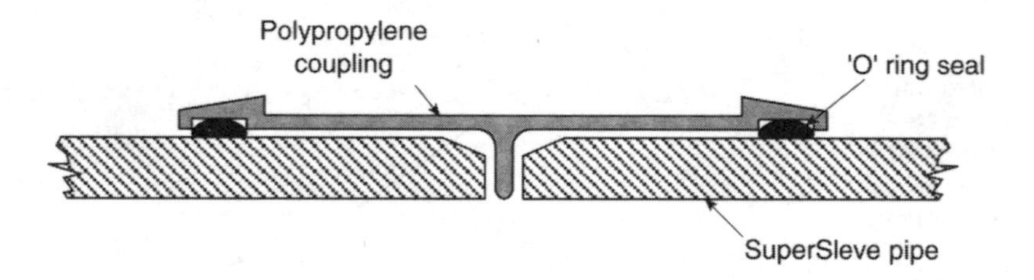

Fig. 11.8 Hepworth 'SuperSleve' joint

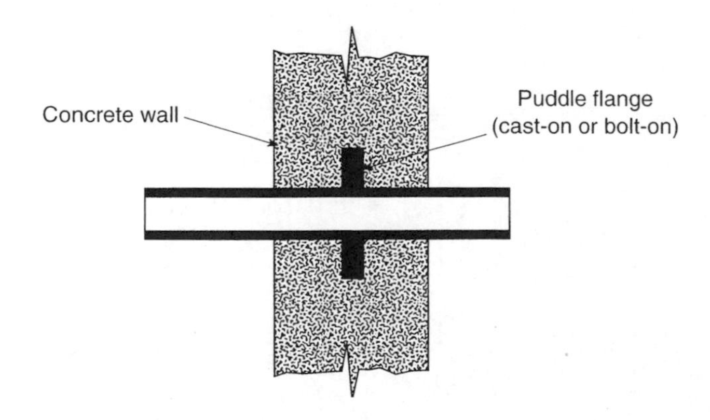

Fig. 11.9 Puddle flange

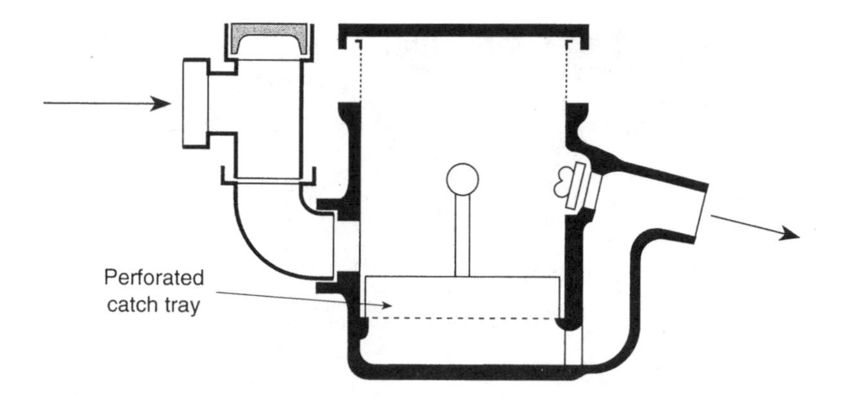

Fig. 11.10 Grease trap

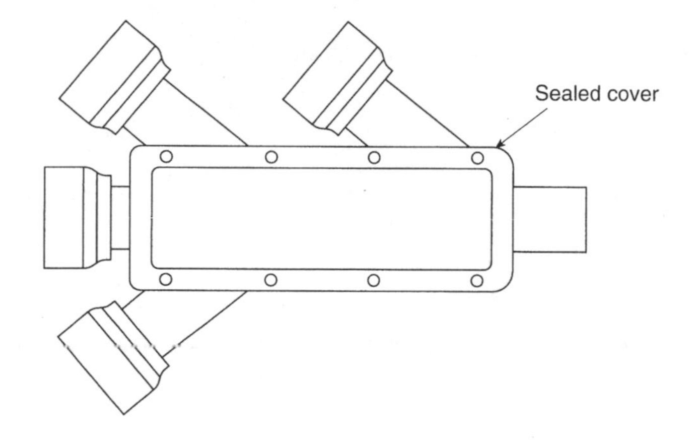

Fig. 11.11 Cast iron inspection fitting

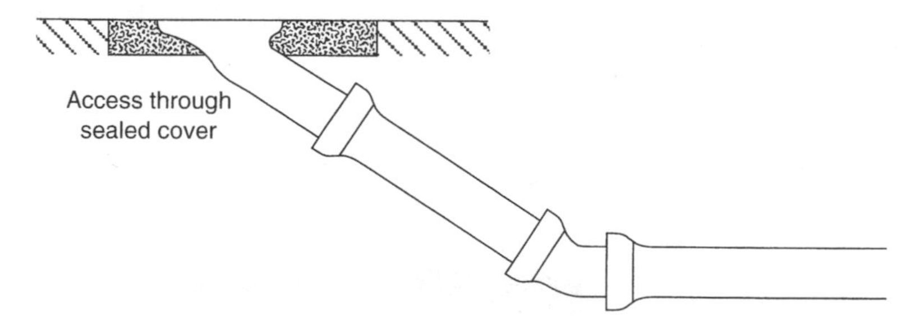

Fig. 11.12 Rodding eye

Now try and answer these questions

1. A ridged joint on a cast iron drainpipe is made by pouring __________ into the socket.

   a cement grout
   b liquefied clay
   c iron cement
   d molten lead.

2. UPVC drainpipe and fittings are normally coloured

   a grey
   b brown
   c blue
   d green.

3. Because of its strength cast iron drainage systems are permissible when laid beneath

   a highways
   b footpaths
   c buildings
   d playing fields.

4. The approved document for the installation of drainage, in England, within the Building Regulations 1985 is

   a Document A
   b Document L
   c Document C
   d Document H.

5. Centrifugal spun cast iron pipe is manufactured to

   a BS 1211
   b BS 1143
   c BS 78
   d BS 1192.

6. A system of below-ground drainage where both foul and surface water are connected to a common drainpipe is known as a

   a one-pipe system
   b combined system
   c separate system
   d partially combined system.

7. When setting up the gradient of a length of drainpipe the __________ of the pipe is used to achieve the correct level.

   a collar
   b spigot
   c soffit
   d invert.

8. A public sewer is the responsibility of the

   a individual users
   b central government
   c Local Authority/water undertaking
   d County Council.

9. A drain fitting which has a water seal and is used to connect a waste discharge pipe to a below-ground drain is referred to as a

   a trapped gully
   b rest bend
   c rainwater shoe
   d anti-flooding valve.

10. A rodding eye is a means of access to a drain which is located at the ________ of the drain.

    a base
    b junction
    c head
    d centre.

11. Where clay drainpipes are to convey sewage with high levels of chemical substances, the pipes should be manufactured to

a BS 1143
b BS 1196
c BS 1152
d BS 1162.

12. UPVC drainpipes are unsuitable where long discharges of __________ are expected.

a rainwater
b high levels of grease
c foul sewage
d hot waste.

13. When vitrified clay pipes are to be laid the trench bedding should be of

a Class O
b Class A
c Class B
d Class D.

14. Where the slope of the land is greater than the gradient of the drain, the gradient is maintained by using ________ manholes.

a inspection
b pre-cast
c backdrop
d inverted.

15. UPVC drainware manufactured to BS 4660: 1971 are pipes and fittings of

a 350 mm diameter
b 75 mm diameter
c 150 mm diameter
d 100 mm diameter.

16. A drainage system which has a pipe system for foul sewage and a pipe for surface water is referred to as a

a combined system
b one -pipe system
c separate system
d partially combined system.

17. A sewer which serves several properties and is laid within those properties is referred to as a

a private sewer
b public sewer
c combined sewer
d utility sewer.

18. The rate of fall of a length of drainpipe is referred to as the

a gradient
b inclination
c drop
d velocity.

19. The preferred method of testing foul drains is to apply a

a drop test
b air test
c profile test
d water test.

20. The responsibility for a private sewer lies with the

a local government
b Water Authority
c central government
d property owners.

21. Where surface water from a large car park or garage is to discharge into a drainage system, the water must first pass through a

a filter screen
b garage gully
c petrol interceptor
d yard gully.

22. Where there is a strong possibility that 'back-flow' will occur from a drainage system, a device called ________ should be installed.

a a catch tank
b an interceptor trap
c a pressure-relief valve
d an anti-flooding valve.

23. Rainwater drop pipes are connected to a surface water drain by means of a

a back inlet gully
b rainwater shoe
c soak-away
d grease trap.

24. A caulked joint on cast iron drainpipe is classified as a

a push-fit joint
b hot joint
c flexible joint
d ridged joint.

25. When cast iron pipe passes through concrete walls, etc., a flexible joint referred to as a __________ is applied on either side of the wall.

a push-fit joint
b victaulic joint
c caulked joint
d flanged joint.

26. When laying drainpipes with a Class B bedding, the minimum depth of granular material is to be

a 50 mm
b 75 mm
c 150 mm
d 100 mm.

27. A drain connecting a single building to a public sewer is called a

a private drain
b public drain
c public sewer
d private sewer.

28. If a cast iron drainpipe passes through a concrete basement wall a device known as a __________ is attached to the pipe before being cast into the concrete.

a DPC
b slip-on flange
c puddle flange
d integral flange.

29. Applying Maguire's rule to a 100 mm drainpipe 40 m long, the total fall would be

a 0.5 m
b 10 m
c 0.1 m
d 1 metre.

30. Applying Maguire's rule to a length of 100 mm diameter drainpipe 8 m long, the total fall would be

a 200 mm
b 0.5 m
c 400 mm
d 0.25 m.

31. Which one of the possible answers listed is a means of access to a drain or sewer?

a garage gully
b manhole
c drain plug
d grease trap.

32. When testing drains for soundness, one method of sealing off a section of drain is to insert a

a cloth bung
b clay bung
c paper bung
d drain plug.

33. A drain fitting called a __________ should be used to connect waste pipes from large kitchens to drainage systems.

a petrol interceptor
b yard gully
c grease trap
d rainwater shoe.

34. A type of clay pipe which gives a reduction in length-to-weight ratio and strength has a trade name of

a SuperSleve
b Hepsleve
c Quickfit
d Mighticlay.

35. The maximum distance between manholes on sewage and drainage systems is

a 40 m
b 25 m
c 45 m
d 90 m.

36. A quick and reliable method of joining drain pipes and fittings is referred to as a

a caulked joint
b push-fit joint
c flanged joint
d compression joint.

37. A drain fitting which minimizes the chance of blockages when connecting waste discharge pipes is called a

a garage gully
b back inlet gully
c yard gully
d rainwater shoe.

38. Drains being tested by a water test should be subjected to a __________ head of water.

a 1.5 m
b 0.5 m
c 2.5 m
d 1.0 m.

39. A drainpipe 25 m in length and 150 mm in diameter should have a total fall of

a 1000 mm
b 250 mm
c 417.5 mm
d 517.5 mm.

40. A self-cleansing velocity in drainage terms is

a 1m/min
b 2m/min
c 2m/s
d 1m/s.

41. One advantage of the combined drainage system is that it is impossible to connect a soil discharge pipe to a

a back-drop manhole
b soak-away
c surface water drain
d Class B bedding.

42. Where a drain changes direction it is desirable to install

a extra bedding
b a petrol interceptor
c a manhole
d an anti-flooding valve.

43. Flexible (push-fit) joints will allow up to __________ of deflection before the joint fails.

a 5°
b 10°
c 15°
d 20°.

44. An inspection chamber is a means of access to a drain which is less than ____ in depth.

a 0.5 m
b 1.5 m
c 20 m
d 1.0 m.

45. A manhole is an access to drains and sewers which are greater than ___ in depth.

a 0.5 m
b 1.5 m
c 2.0 m
d 1.0 m.

Now check your answers from the grid.

Q 1; d	Q 10; c	Q 19; b	Q 28; c	Q 37; b
Q 2; b	Q 11; a	Q 20; d	Q 29; d	Q 38; a
Q 3; c	Q 12; d	Q 21; c	Q 30; a	Q 39; c
Q 4; d	Q 13; b	Q 22; d	Q 31; b	Q 40; d
Q 5; a	Q 14; c	Q 23; b	Q 32; d	Q 41; c
Q 6; b	Q 15; d	Q 24; d	Q 33; c	Q 42; c
Q 7; d	Q 16; c	Q 25; b	Q 34; a	Q 43; a
Q 8; c	Q 17; a	Q 26; d	Q 35; d	Q 44; d
Q 9; a	Q 18; a	Q 27; a	Q 36; b	Q 45; d

12

SANITARY PIPEWORK

To tackle this chapter you will need to know:

- the type of discharge systems installed;
- the means of ventilating discharge systems;
- the materials used in discharge systems;
- the effects of pressure fluctuations;
- the type of traps used on sanitary appliances;
- the methods by which trap seals are lost.

GLOSSARY OF TERMS

Access cover – a permanent means of access to internal pipework. It may be found on the pipe or the pipe fittings, and provides access for testing, inspection and cleaning.
Access for maintenance – within the system there should be adequate access covers or cleaning eyes to permit cleaning or clearing of the entire system.
Air admittance valves (AAV) – automatically operated valves which permit air in to a system when sanitary appliances discharge into a stack and cause a negative pressure. They can be used as a substitute for a dry vent stack when termination of the latter would be difficult. They should only be installed with reference to the *Building Regulations: Document H* in England and *Part M Building Standards* in Scotland and the manufacturer's instructions. All AAVs must have a British Board of Agreement Certificate.
Air test – a test that is achieved by fully charging all appliance traps with water on the system and sealing the bottom and top of the stack drain bungs. A rubber tube is then passed through a WC trap and connected to a test tee which is in turn connected to a manometer and air pump. The pressure in the system is elevated to 38 mm water gauge for a period of not less than 3 minutes.
Anti-siphon trap – a trap that has a device which enables air to enter the system when siphonic tendencies occur during discharge thus preventing a loss of seal.
Borosilicate glass – a system of glass tubes and fittings with proprietary joints used in laboratory waste discharge systems to convey chemical or radiation waste.
Bottle trap – a type of trap where the division between the inlet and outlet is formed either by a dip tube or a vane within the trap body. Usually the bottom half of the trap can be removed for cleaning purposes.
Branch discharge pipe – a section of pipe which connects a sanitary appliance to a discharge stack.
Branch ventilating pipe – a section of pipe used to provide permanent ventilation to a branch discharge pipe thus providing permanent protection from siphonic tendencies in the system.
BS 5572 1994 – the British Standard which replaced BS: CP304, which was a 'Code of Practice', and is the current standard for the installation of sanitary pipework.
Criterion for satisfactory service – This is a term which dictates the percentage of time when the design criteria for discharge flow loading must not be exceeded.
Crown of trap – the topmost point of the inside of a trap outlet.
Depth of water seal – the water held in a trap is the means of preventing free air flow through the trap. The depth of seal is measured from flood level of the outlet to the lowest point of the method of seal.
Discharge – the rate of discharge from appliances, and varies from type to type. Typical values are given in British Standards and Guides and are the basis for the design of the system.
Discharge pipe – a pipe which carries the discharge from a sanitary appliance.
Gradients (of branch discharge pipes) – the gradient of a branch discharge pipe should be such that it prevents siphonic tendencies and provides a self-cleansing discharge to take place. The minimum recommended gradient is 1° to 1.25° which is equivalent to 18 mm/m or 22 mm/m respectively.
Lengths (of branch discharge pipes) – branch discharge pipes should be kept to the shortest practicable length, with due regard given to branch pipes serving wash basins and urinal bowls. This has the effect of reducing both self-siphonage, and the build-up of waste deposits.

Loss of Seal – can be caused by the following:

i induced siphonage
ii self-siphonage
iii capillary action
iv evaporation
v momentum
vi back pressure.

Induced siphonage – the loss of seal in a trap when a full-bore flow occurs in a main branch discharge pipe, creating a negative pressure in the discharge pipe from the trap.
Self-siphonage – the loss of trap seal through self-siphonage is usually towards the end of a discharge particularly where a long discharge pipe is present. A full-bore flow will tend to create a negative pressure zone at the back of the trap, thus causing atmospheric pressure to push the trap seal into the discharge pipe.
Capillary action – seal loss through capillary action can be caused by materials such as string, cloth, toilet tissue or sanitary towels being left behind after a discharge both in contact with the liquid seal and trailing into the discharge pipe. Seal loss can be either rapid or slow.
Evaporation – Seal loss through evaporation is most likely to occur when the period of time between discharges is lengthy. The liquid simply evaporates into the internal atmosphere.
Momentum – loss of seal by means of momentum is caused by severe pressure fluctuations created by high-velocity wind flowing across the top of the dry vent stack. The high waves created in the trap liquid allow sufficient volume to flow into the discharge pipe to cause a loss of seal. Termination of the stack will have a bearing on this action.
Back pressure – this type of loss of seal usually occurs at appliances situated at the base of a stack. When high rates of discharge take place – high pressures are present in the lower section of the stack – this results in foul air being blown through the trap and in severe cases the liquid is thrown clear of the appliance.
Resealing trap – a trap that has an inbuilt capacity to retain sufficient liquid to reseal the trap when a siphonic action has occurred.
Resistance to blockage – a discharge system should be designed so that the risk of blockage is limited to an absolute minimum.
Sanitary pipework – a system of discharge pipes, which may or may not have vent pipes, that is connected to a drainage system.
Single stack system – a system where the discharge stack is large enough to prevent pressure differentials without the use of ventilating pipes.
Stack – a main vertical discharge or ventilating pipe to which all the branch discharge pipes are connected. The size of the stack is dependent on the design load.
Stack vent (dry vent) – this is often referred to as the length of pipe above the highest branch discharge pipe entering the wet stack which extends above the highest opening to the building terminating in an end open to atmosphere.
Stub stack – stack constructed from a short straight 100 mm discharge stack with the top sealed. This would normally be an access cap for cleansing purposes. It may be used to connect a limited number of appliances to a drain, provided that the centre of the WC branch connection is no more than 1.5 m from the invert of the drain and the centre of the topmost waste branch connection is no more than 2.5 m from the invert of the drain.
Testing – all parts of the system should be provided with access to permit testing of the entire system. Systems are tested with air or water.
Trap – either a pipe fitting or part of a sanitary appliance that retains liquid to prevent the ingress of foul air into a building.
Tubular traps – traps manufactured from pipe or moulded in the shape of circular pipe. The sizes of tubular traps should not be less than the dimensions found in Table 12. 1.
Type of materials (metals) – the following list of metals is generally acceptable in the construction of sanitary pipework systems

i cast iron
ii copper
iii galvanized steel
iv lead
v stainless steel

Table12.1 Minimum sizes of tubular traps (BS 5572: 1994)

Type of appliance	Size of trap (mm)	Type of appliance	Size of trap (mm)
Wash basin	32		
Bidet	32		
Sink	40		
Bath	40		
Shower bath tray	40		
		Urinal (bowl)	40
		Urinal (1 to 7 stalls or slab of equivalent length)[1]	65
Drinking fountain	32	Food waste disposal unit (domestic)	40
Bar well	32	Food waste disposal unit (industrial type)	50
		Sanitary towel macerator	40

1 Where there are more than seven stalls or a slab of equivalent length in one range, more than one outlet should be provided.

Cast iron – Cast iron pipe and fittings are manufactured by pouring molten iron into sand-casts. The castings are then immersed in a hot bath of a bituminous mixture called 'Dr Angus Smith's solution', which gives the material an extended protection against corrosion. Joints are made by pouring molten lead into a prepared spigot and socket type joint. The cooled lead is then 'caulked' by hammer and caulking tools. This is a traditional material which nowadays is only used when specified.
Copper – Copper pipe installations are normally found within buildings in specially constructed service ducts. Jointing is either by forming joints and bronze welding them or alternatively by using specially manufactured brazing fittings.
Galvanized steel – This type of system is suited mostly to buildings where the pipework is repeated. Sections of low-carbon steel pipe are prefabricated and dipped in baths of molten zinc, which forms the galvanizing coat, before being delivered to site for installation. Jointing is of the spigot and socket type filled with a proprietary jointing medium or alternatively small-diameter copper tubes are made into female threaded connections welded into the branch and stack pipes.
Lead – Lead is the forerunner of all pipe materials traditionally used by the plumber. The joints were initially made by wiping solder around the joints (wiped joints). More recently joints are made by lead-welding techniques. Lead and cast iron pipes were often combined in pipework installations.
Stainless steel – This material is for special installations and only used in very high specification building work. The jointing of stainless steel pipe is by the MAG welding technique.
Types of materials (plastics) – The following list of plastic materials is generally acceptable in the construction of sanitary pipework systems

- i acrylonitrile butadiene styrene (ABS)
- ii high-density polyethylene (HDPE)
- iii modified unplasticized polyvinylchloride (MUPVC)
- iv unplasticized polyvinylchloride (UPVC)
- v polypropylene (PP).

Acrylonitrile butadiene styrene – This thermoplastic material is used in the manufacture of soil and waste pipes. The method of jointing is by applying a solvent cement to the two surfaces being jointed, i.e. the spigot and socket surfaces. The solvent cement dissolves the surfaces of the joint, which fuse together. The joint sets within 10 minutes, but may take 24 hours before the joint becomes fully cured.

High-density polyethylene – This material is also a thermoplastic, which is used in the manufacture of specialized waste systems. These systems are joined together using heated tools, which fuse the two surfaces to produce a single mass of material, a process referred to as 'fusion welding'.

Modified unplasticized polyvinylchloride – Another member of the thermoplastic family of plastics is used to manufacture general-purpose soil and waste pipe and fittings. It is an adaptable and easily moulded material. Jointing methods include push-fit 'O' ring, compression and solvent welded joints. A wide range of products gives great adaptability and ease of installation. The product is prone to ultraviolet degradation and advice should be sought from the manufacturer.

Unplasticized Polyvinylchloride – A very similar material to MUPVC in its application.

Polypropylene – Like polyethylene, this material is used to manufacture specialist waste systems. Jointing is by the fusion welding process using proprietary tools.

Ventilated stack system – a system used where close grouping of appliances makes it possible to use branch discharge pipes without the necessity of having branch ventilation pipes. The discharge stack is cross-connected to the ventilating stack.

Ventilated System – a system used where large numbers of sanitary appliances are grouped in ranges or are widely dispersed. The trap seals are protected by providing a ventilation stack with a branch ventilating pipe connected and extending the discharge stack to atmosphere.

Ventilating pipe – an integral part of the system which limits pressure differences when there are discharges taking place, thus eliminating siphonic tendencies.

Water test – A water test should not be applied to the whole system but simply to the lower sections of the system mainly at risk of flooding. This is normally up to the flood level of the lowest sanitary appliance of the system. A drain plug is inserted in the drain at the nearest manhole and filled with water to the lowest appliance, provided that the static head does not exceed 6 m.

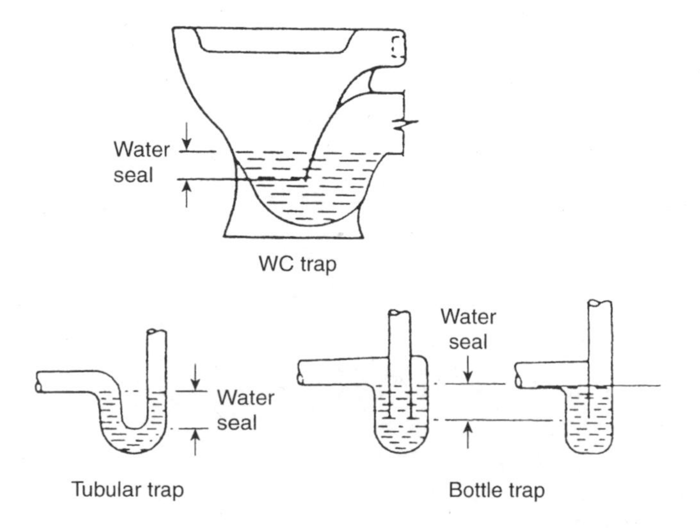

Fig 12.1 Trap types

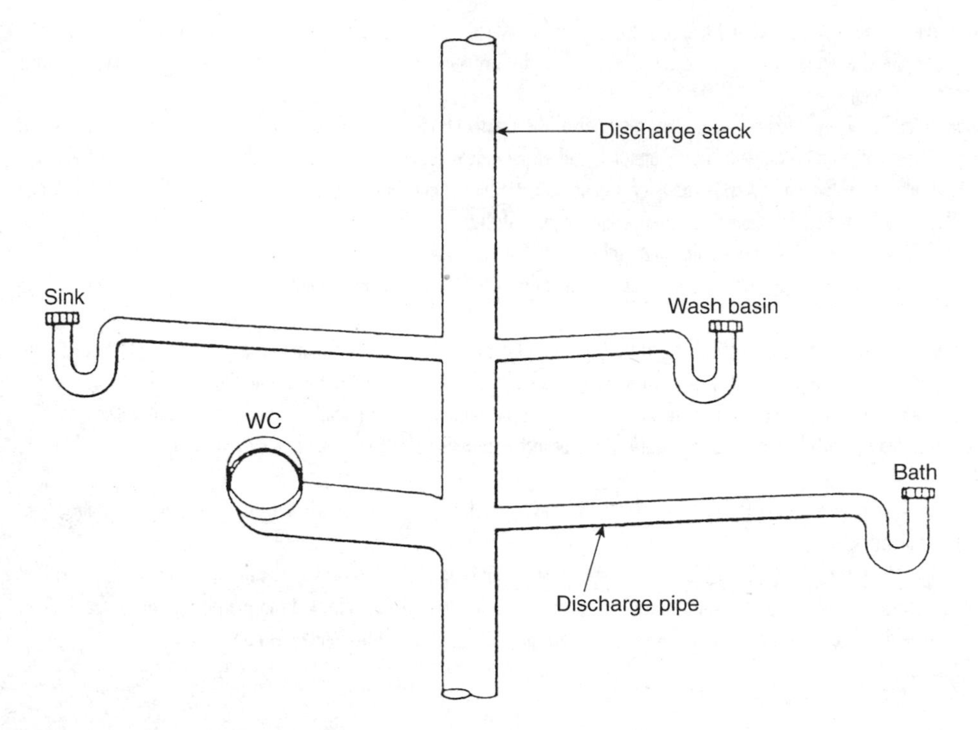

Fig 12.2 Single stack

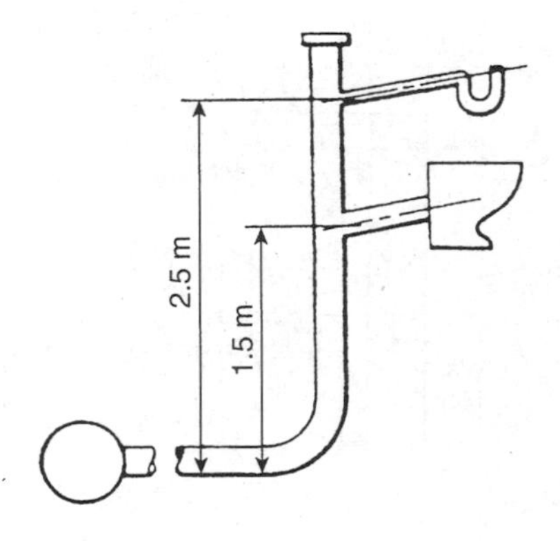

Fig 12.3 Stub stack

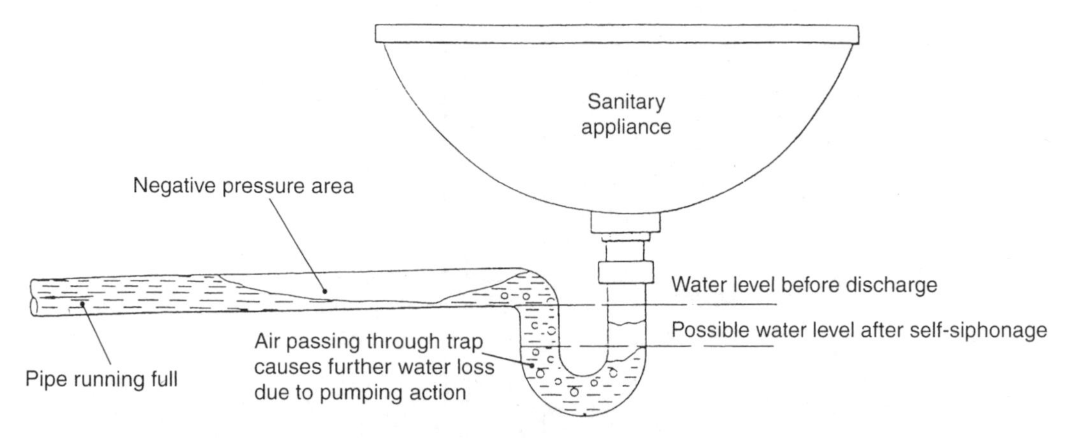

Self-siphonage (at the end of an appliance discharge)

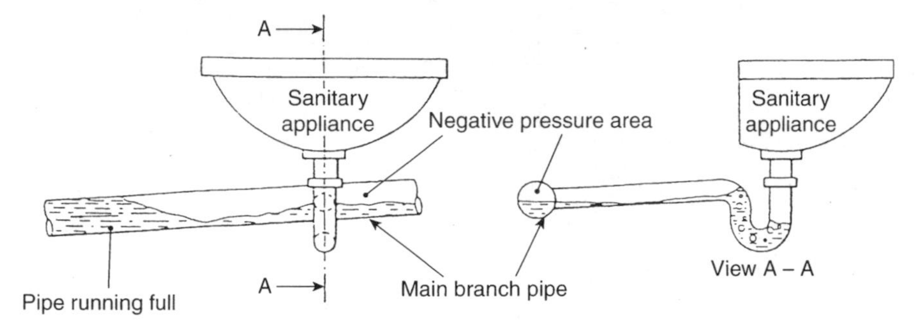

Induced siphonage (due to full bore flow in a main branch discharge pipe)

Fig 12.4 Seal loss due to flow in branch pipes

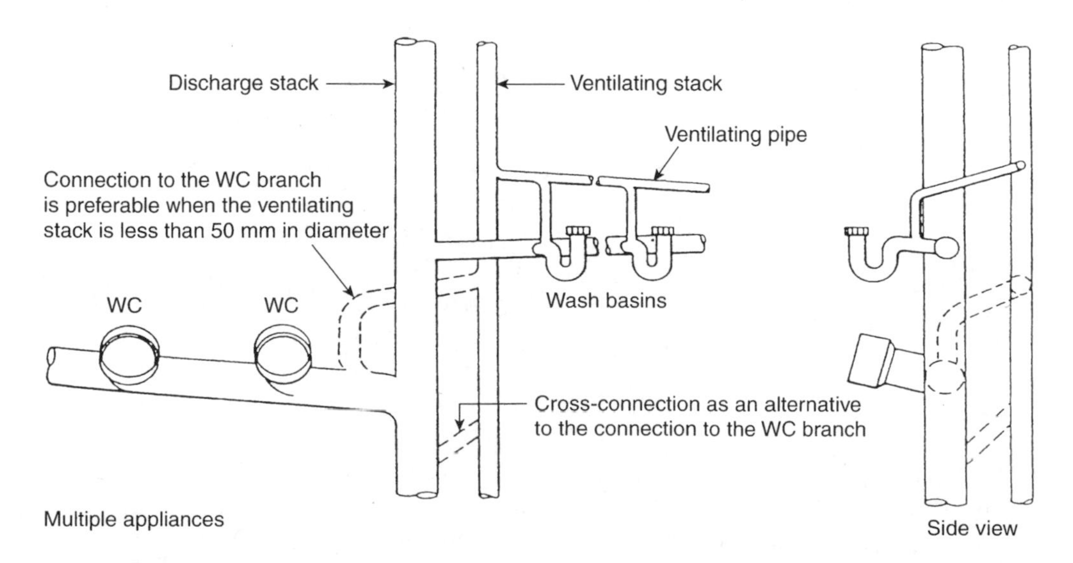

Fig 12.5 Ventilated system

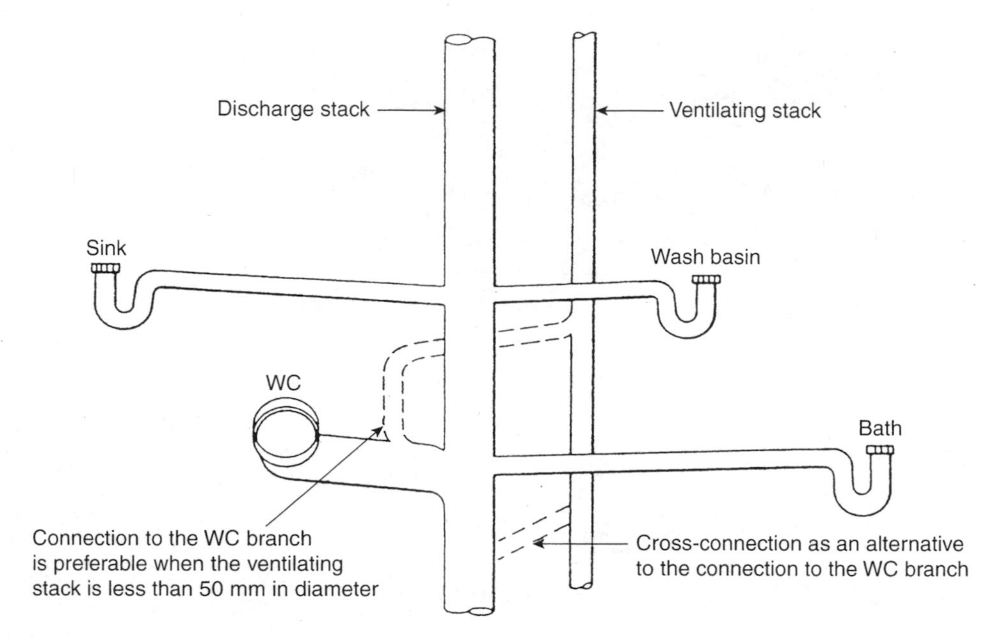

Fig 12.6 Ventilated stack system

Now try and answer these questions

1. When the seal of a trap is lost during a time when a full bore flow occurs in a main discharge pipe, this is referred to as
   - a self-siphonage
   - b back pressure
   - c induced siphonage
   - d momentum.

2. One of the causes of loss of seal in traps is when __________ is present in the discharge pipe.
   - a a negative pressure
   - b high pressure
   - c a positive pressure
   - d gas pressure.

3. The purpose of a trap on a sanitary fitting is to
   - a prevent the egress of foul air
   - b connect it to a discharge pipe
   - c connect it to a ventilating stack
   - d prevent the ingress of foul air.

4. The main vertical discharge or ventilating pipe is referred to as the
   - a stack
   - b float
   - c vent
   - d riser.

5. The size of a stack pipe is dependent upon the __________ of the system.
   - a total height
   - b design load
   - c owner
   - d materials.

6. Sanitary appliances which are used infrequently are prone to loss of seal in their traps through

a momentum
b self-siphonage
c back pressure
d evaporation.

7. In laboratory installations sanitary pipework is often installed with ______ glass.

a silicon
b crystal
c borosilicate
d gyposilicate.

8. When high rates of discharge occur at the base of a stack, high pressure can cause loss of seal by

a back pressure
b base pressure
c self-siphonage
d induced siphonage.

9. The choice of termination position of a stack vent can help prevent loss of seal in traps by

a back pressure
b momentum
c capillary action
d evaporation.

10. When the discharge pipe connected to a trap is particularly long, the trap seal can be lost by means of

a capillary action
b momentum
c induced siphonage
d self-siphonage

11. The pipe connecting a sanitary appliance and the vertical stack is referred to as a

a branch ventilating pipe
b branch discharge pipe
c sanitary pipe
d vent pipe.

12. The type of trap incorporating a device which admits air into the system when siphonic tendencies occur is called a

a bottle trap
b Grevak trap
c anti-siphon trap
d knot trap.

13. After siphonic action has occurred a trap which has an inbuilt capacity to reseal the trap is referred to as a

a resealing trap
b anti-siphon trap
c Grevak trap
d bottle trap.

14. The current standard for the installation of sanitary pipework is

a BS 6644: 1992
b BS 5440: 1989
c BS 5572: 1994
d BS: CP304: 1969.

15. An access cover is a permanent means to access the internal pipework for

a venting; painting; inspecting
b venting; painting; testing
c testing; painting; cleaning
d testing; inspecting; cleaning.

16. Sanitary pipework systems manufactured using galvanized steel pipe are firstly pre-fabricated and then

a sprayed with oxide paint
b dipped in molten zinc
c coated in plastic
d dipped in bitumen.

17. A cast iron pipe joint is made by running molten lead into the socket. This method of jointing is called

a an 'O' ring joint
b a compression joint
c a caulked joint
d a hot joint.

18. Cast iron pipe and fittings are dipped into a bituminous solution to provide protection against
    - a corrosion
    - b blockage
    - c leakage
    - d adhesion.

19. Sanitary pipework systems constructed with copper tubes may be jointed using the __________ technique.
    - a fusion welded
    - b bronze welded
    - c extrusion
    - d intrusion.

20. The earliest metal used for sanitary pipework was
    - a asbestos
    - b wrought iron
    - c plastic
    - d lead.

21. The most suitable types of plastics for the manufacture of soil and waste pipes are
    - a foam plastics
    - b thermosetting plastics
    - c thermoplastics
    - d thermodynamic plastics.

22. When using acrylonitrile butadiene styrene pipes and fittings, the most favoured method of joint is by using a
    - a solvent cement
    - b push-fit socket
    - c flanged type
    - d friction type.

23. Polyethylene or polypropylene pipes are ______ together when forming a joint.
    - a crimped
    - b fusion welded
    - c flanged
    - d solvent cemented.

24. Soil systems constructed with MUPVC pipes and fittings usually employ a jointing technique referred to as
    - a push-fit 'O' ring
    - b fusion welding
    - c grab-ring jointing
    - d slip-ring jointing

25. Polyvinylchloride products should be protected from solar radiation to prevent
    - a creep
    - b capillary
    - c infrared warming
    - d ultraviolet degradation.

26. Solvent cemented joints on ABS plastic pipe and fittings would normally cure in
    - a 24 minutes
    - b 10 hours
    - c 24 hours
    - d 10 minutes.

27. Solvent cemented joints on ABS plastic pipe and fittings would normally set in
    - a 24 minutes
    - b 10 hours
    - c 24 hours
    - d 10 minutes.

28. In the stub duct system the maximum height above the invert of a drain to the center of a WC branch is
    - a 2.5 m
    - b 1.5 m
    - c 15.0 m
    - d 5.0 m.

29. An air admittance valve may only be used to ventilate a stack if the valve possesses a
    - a British Standards Institute Certificate
    - b British Board of Agrément Certificate
    - c British Plumbers Certificate
    - d ISO 9000 Certificate.

30. When planning the location of a sanitary pipework stack, consideration should be given to the __________ of discharge branches.

  a length
  b fixing
  c colour
  d jointing.

31. The length of a discharge branch pipe can have an influence on the amount of __________ through self-siphonage.

  a cleansing
  b blockage
  c loss of seal
  d leakage.

32. The section of the *Building Regulations* which stipulates where air admittance valves may be installed is

  a Document A
  b Document F
  c Document L
  d Document H.

33. The minimum recommended gradient for branch discharge pipes is between

  a 10° and 12°
  b 5° and 8°
  c 1° and 1.25°
  d 2° and 3°.

34. The end cap of a stub stack system provides access for

  a cleaning
  b ventilating
  c future connections
  d painting.

35. The minimum size of tubular trap that can be connected to a wash basin is

  a 54 mm
  b 32 mm
  c 42 mm
  d 40 mm.

36. The minimum size of tubular trap that can be connected to a shower bath tray is

  a 54 mm
  b 32 mm
  c 42 mm
  d 40 mm.

37. When a slab urinal is installed containing more than seven stalls __________ outlet(s) should be provided.

  a one
  b two
  c three
  d four.

38. The ventilated system is used when __________ of sanitary appliances are grouped in ranges.

  a large numbers
  b different sizes
  c different colours
  d similar types.

39. The single-stack system is where the discharge stack is large enough to prevent

  a frost damage
  b pressure differentials
  c multiple connections
  d momentum occuring.

40. Where corrosive chemicals are to be conveyed to waste the most suitable material to use would be

  a copper
  b asbestos
  c UPVC
  d borosilicate glass.

41. An air test on a sanitary pipework system should be carried out by increasing the internal air pressure to

  a 100 mm
  b 120 mm
  c 38 mm
  d 28 mm.

42. The duration of an air test for sanitary pipework should be for a period of not less than

   a 2 min
   b 3 min
   c 5 min
   d 10 min.

43. A water test on sanitary pipework is carried out to the flood level of the __________ sanitary appliance.

   a lowest
   b highest
   c smallest
   d widest.

44. The main vertical discharge of ventilating pipe is referred to as a

   a riser
   b dropper
   c stack
   d float.

45. A discharge system should be designed so that the risk of __________ is limited to an absolute minimum.

   a blockage
   b overuse
   c interference
   d over-pricing.

Now check your answers from the grid.

Q 1; c	Q 10; d	Q 19; b	Q 28; b	Q 37; b
Q 2; a	Q 11; b	Q 20; d	Q 29; b	Q 38; a
Q 3; d	Q 12; c	Q 21; c	Q 30; a	Q 39; b
Q 4; a	Q 13; a	Q 22; a	Q 31; c	Q 40; d
Q 5; b	Q 14; c	Q 23; b	Q 32; d	Q 41; c
Q 6; d	Q 15; d	Q 24; a	Q 33; d	Q 42; b
Q 7; c	Q 16; b	Q 25; d	Q 34; a	Q 43; a
Q 8; a	Q 17; c	Q 26; c	Q 35; b	Q 44; c
Q 9; b	Q 18; a	Q 27; d	Q 36; d	Q 45; a

13

SHEET LEAD FLASHINGS

To tackle this chapter you will need to know:

- the type and codes of sheet lead used in building;
- the methods of securing and fixing sheet lead;
- methods of jointing and providing for thermal movement;
- the specialist tools required for working sheet lead;
- the different types of flashings.

GLOSSARY OF TERMS

Bossing stick – the preferred implement for moving lead to form complex shapes in sheet lead. It consists of a body carved from wood, having no sharp edges and a handle formed from the same piece of wood. Due to the lack of sharp edges it causes none or little damage to the lead on impact. Bossing sticks are available made from boxwood or plastic.
British Standard 1178: 1982 – the British Standard, *Milled Sheet Lead for Building Process* sets out the criteria for the production of lead sheet in terms of chemical composition and thickness. It is designed to provide quality and consistency.

BS1178: 1982 Code No	Thickness (mm)	Weight (kg/m^2)	Colour Code
3	1.32	14.97	Green
4	1.80	20.41	Blue
5	2.24	25.40	Red
6	2.65	30.05	Black
7	3.15	35.72	White
8	3.55	40.26	Orange

Chase wedge – a tool that resembles a broad chisel manufactured from boxwood or plastic. A metallic ring protects the shaft end. The drift is used to drive lead into sharp corners. Caution should always be used with this tool as considerable damage can result from misuse.
Chemical corrosion – sheet lead should never be laid directly on to a wood substructure made of hardwood, particularly oak. Condensation forming on the underside of the lead will leech out chemicals within the oak to produce acids that will attack and break down the lead.
Chimney apron – the flashing used to weather the front aspect of a chimney and provide an 'apron' cover for water to run off the chimney. Traditionally it is made by the bossing technique from Code 4 or 5 lead. In recent times lead welding has become popular. The apron consists of the front apron, an upstand and two return ends taking it 100 mm up the sides of the chimney. It is wedge fixed into the wall. The apron is the first component of a chimney flashing to be fixed.
Chute outlet – an outlet formed from a single piece of sheet lead. It consists of two upright sides and a sole. It is a means of draining a flat roof contained by parapet walls. It passes through the wall to discharge over a hopper head mounted on top of a rainwater pipe.
Clips – a method of securing lead work to buildings. Clips or cleats may be produced from either copper, stainless steel or lead. In the case of copper the clips should be no less than 50 mm wide and no less than 0.6 mm thick. Stainless steel clips should be cut from austenitic stainless steel no less than 50 mm wide and 0.38 mm thick. Because of the malleability of lead, lead clips should only be considered in very sheltered sites and from a code equal to or greater than the one being fixed.
Continuous step flashing – a type of flashing that is particularly suited to brickwork construction, where there is a continuous seam and brick thickness. It is used to weather pitched roofs to abutment walls and chimney stacks. Strips of sheet lead 150 mm wide and cut to suitable lengths (max 1.5 m) are placed against the abutment. A water line 65 mm up the strip determines the extent of the horizontal step measurement; a 25 mm 'turn-in' is added before the waste is cut off. The step flashing is then wedged into the wall covering the soaker upstands. Code 4 lead is suitable for this application.

Corrosion resistance – lead is one of the most corrosion-resistant metals, although when lead is exposed to the atmosphere the bright metallic lustre is attacked by the oxygen, carbon dioxide and water vapour present in air. The substances react with the lead to produce a film on the lead surface. This film – lead oxide – will then protect the lead beneath. The lead oxide film has a whitish grey appearance, which builds up with maturity.

Cover flashing – strips of sheet lead which are wedged into an abutment wall and dressed down over the upstand of a flat roof, for instance, whether it be built-up felt or lead. The cover flashing should give a minimum of 75 mm cover over the upstand. This prevents water penetration through capillary attraction. The length of a cover flashing should not exceed 1.5 m. Individual lengths of flashing should have a minimum 100 mm lap to ensure full weathering.

Damp proof course (DPC) – sheet lead has long been used as a damp proof course in building construction. Although modern materials have superseded lead for general DPCs, lead is still viable where complex shapes and applications are required. Bossing or welding may be used to produce these components. Code 3 lead is suitable for this application.

Dresser – the heavyweight of lead tools. It has a broad flat surface for flattening out sheet lead or for finishing (dressing) the lead to a surface. It is not really suited to the bossing of lead. It is traditionally manufactured from boxwood and in recent times from plastic.

Drift plate – a sheet of steel plate, which can be inserted between two overlapping pieces of lead. It enables the top sheet to be worked easily by preventing friction holding the two sheets together.

Electrolytic corrosion – from experience it has been found that there is no significant risk from this type of corrosion when lead is in contact with copper, stainless steel, iron, zinc or aluminium.

Galena – The ore from which lead is produced, containing 86.6% pure lead and 13.4% sulphur. It has a cubic crystalline structure. It is very lustrous with a dark grey appearance. It is mined in many parts of the world.

Gutter back – a flashing that covers a wood-lined gutter formed at the back of chimneys or abutment walls. It is manufactured from Code 4 and 5 lead by the bossing or welding technique. The gutter has an upstand to the abutment, a gutter sole (150 mm), an extension up the roof slope which is overlapped by the roof covering and two returns down the sides of the chimney. This is the last chimney flashing component to be applied.

Hip roll – this is a similar application to ridge roll (see below), but in this instance the solid wood roll is mounted on the hip rafter on a hipped pitched roof.

Lap joint – a simple jointing technique whereby one piece of lead overlaps another – undercloak and overcloak. It also provides a valuable expansion joint which prevents failure by cracking. The amount of lap is dependent on the angle of lay.

Lead – this is the softest of the common metals used in building. It has high properties of ductility, malleability and corrosion resistance. One of its most endearing properties is its ability to be worked at normal temperatures without having to be annealed, as work hardening is not appreciable.

Lead bossing – a traditional craft associated with plumbers; in fact the name plumber is derived from the Latin word *plumbum*, which means lead. It is a technique whereby the great ductility and malleability of lead is exploited to form simple and complex shapes from flat sheets of lead. Skills involved include the ability to move lead from one part of a street to another without tearing, cracking or losing any of the thickness of the sheet. The plumber uses a selection of tools, either manufactured or self-styled, to achieve his goal.

Lead sheet sizes – sheet lead is available in rolls ranging from 150 mm to 600 mm in width. The widths go up in 30 mm increments. The lengths of these rolls are either 3 m or 6 m. A range of other lengths and widths is available on request.

Lead slate – a type of flashing used where a pipe, etc., penetrates a flat or pitched roof. In the case of the former it is integrated with the roof covering and in the case of the latter it interlocks with the slates or tiles. It is made up of a sleeve and a flat sheet. It is possible to boss a slate from a single sheet of lead or alternatively fabricate one from two pieces of lead by lead welding.

Lead welding – a relatively recent skill adopted by the plumber. It requires specialized equipment consisting of compressed gas cylinders (oxygen and distilled acetylene) gas regulators, special blowtorches and tubes and of course the safety requirements necessary for this equipment. This form of welding is referred to as autogenous welding, that is to say the filler rod material is the same as the parent metal (lead). It is a process of fusing two pieces of lead to form one. The technique employed is the 'leftward' method of welding. It is essential for the production of sound joints that all surfaces being welded are scraped down to virgin metal and that the flame used is a 'neutral' flame.

Mallet – a lead-working tool which has a boxwood head and a hickory shaft. The mallet head is conically shaped with rounded ends. It resembles an egg in shape. Mallets are made in various sizes to match the job at hand. This is a versatile tool in its application. It can be used to move lead, support the bossing of corners or for delicate finishing off of a piece.

Milled sheet lead – lead sheet is produced by passing a slab of pure lead, 125 mm thick, through a rolling mill many times until the lead is reduced to the desired thickness. The lead is supported on a flat bed during the rolling process.

Nails – nails used for fixing sheet lead should be of the large round-headed type manufactured from either copper or stainless steel. A nail with a serrated shank is preferred. Recommended diameters for shanks are 3.35 mm for copper and 2.65 mm for the stainless steel type. The minimum length for the shank is 19 mm.
Organic corrosion – consideration should be given to the long-term life of sheet lead in areas prone to lichen growth on roofs. Dead and decaying lichens can combine with rainwater to produce solutions, which will take lead into solution and therefore cause the lead to literally run away. This can be prevented by applying patination oil periodically.
Patination oil – an oil applied to the exposed surface of newly fixed lead sheet. It prevents the formation of a white carbonate forming on the surface in damp conditions. This carbonate will form in irregular patterns, which defeats the aesthetic appearance of the lead. A coating of patination oil will protect the sheet lead's surface until it has had time to produce the protective film of lead oxide.
Ridge roll – a method of weathering the ridge of a pitched roof. A solid wood roll is mounted on the ridge board and lead is dressed around it with a cover flashing extending down over the roof covering. Copper straps held in place by the roll hold down the lead.
Saddle – a flashing which is applied to the apex of two valley or secret gutters, and also where the ridge of a pitched roof is against an abutment wall. It is formed to cover the apex and lap over the two top sections of the gutters of step flashings. It can be wedge fixed into a wall or have a cover flashing to weather it to an abutment. It is manufactured by bossing or lead welding from Code 4 or 5 lead.
Screws – the choice of materials for the manufacture of screws for sheet lead fixing is limited to brass or stainless steel. They should comply with BS 1210 with a minimum length of 19 mm.
Secret gutter – a gutter formed where a pitched roof joins an abutment wall. The gutter is lined with lead pieces of a maximum length of 1.5 m and lapped joints. The upstand to the abutment wall acts as a continuous soaker to the step flashing. Then the roof covering is applied and the lead-lined gutter is concealed from view, hence the name. This gutter discharges into either a box gutter or a fascia gutter. Code 4 lead is suitable for this application.
Setting-in stick – a lead tool used to set the fold lines on the sheet before fabrication commences or alternatively it is used to finish off the work, to ease the lead into its final form. Manufactured from either boxwood or plastic, it consists of a handle and body which has a narrow rounded edge and a broad impact edge on top. It is used by lining the narrow edge on a fold line and striking the stick with a mallet.
Single step flashing – a method of weathering a pitched roof to an abutment wall or chimney stack. It is particularly useful with random build stone structures; although it is equally suited to brickwork. Individual flashings are cut to suit the thickness of the masonry courses which take into account the pitch of the roof. Each flashing has a 25 mm 'turn-in' for wedging into the seams. The step flashings are fixed from the bottom of the pitch upward with a minimum lap of 50 mm. The base of the step flashing should cover the soaker by 65 mm. Code 4 lead is suitable for this application
Soaker – soakers are used where slate or tiled roofs abut a wall or chimney. They are used in conjunction with step flashings. The length of a soaker is determined by the distance between the slate lathes (gauge) plus the lap of the slates/tiles. An additional 25 mm is added to be bent down the lathe to prevent it sliding out from under the slate/tile.

Length = Gauge + Lap + 25 mm

Soakers are cut from rolls of lead 175 mm wide; a 90° fold is made in the lead along its length of 100 mm, thus giving 75 mm against the abutment and 100 mm beneath the slate/tile. Soakers are normally manufactured from Code 3 lead.
Thermal movement – lead has a high coefficient of expansion: 0.000 029 m/°C, and therefore consideration must be made when designing sheet lead work. The major causes of lead failure are oversizing and fixing. The provision of expansion joints is crucial for the longevity of the lead work. The following recommendations are made for maximum lengths for flashings.

BS 1178: 1982 Code No	Thickness (mm)	Uses	Maximum length (m)
3	1.32	Soakers	1
4	1.80	Flashings	1.5
5	2.24	Flashings	1.5

Valley gutter – a method of weathering the joint between two pitched roofs intersecting at angles less than 180°. The valley is lined with soft wood boards of suitable width to provide an adequate cover by the roofing material. A tilt fillet runs parallel to the centre of the valley at a distance of 150 mm minimum on both sides. The lead is laid in lengths of 1.5 m working up from the bottom, and is dressed into the gutter and over the tilt fillets. The outside edges should be turned over to act as water stops. The top of each length should be fixed to the gutter with a double row of nails 50 mm apart and 25 mm between rows. No other fixing is required. The lengths of gutter lining should overlap one another by a minimum of 100 mm. Code 4 lead is suitable for this application.

Valley soakers – an alternative to lead-lined valley gutter. Valley soakers are laid in a similar way as roof slates or tiles. A soaker is laid with every course of slate/tile thereby weathering the joint where the two pitched roofs meet. The finished roof conceals the soakers from view. This method of flashing is less prone to defects suffered by oversized pieces of lead in a valley gutter.

Wedge – a means of securing lead flashings to brick or stone-built structures. The joint or seams between the courses of brick are raked, chiselled or cut out to a minimum depth of 25 mm. The wedge is manufactured by the plumber from 20 mm strips of lead normally cut from the waste material. The strip of lead is folded a number of times dependent upon the thickness of the joint and knocked into a wedge shape with a hammer. A 25 mm 'turn-in' on the flashing is inserted into the joint, then the wedge is driven into the joint with hammer and chase to firmly fix the flashing. A wedge should be placed at 250 mm intervals where appropriate.

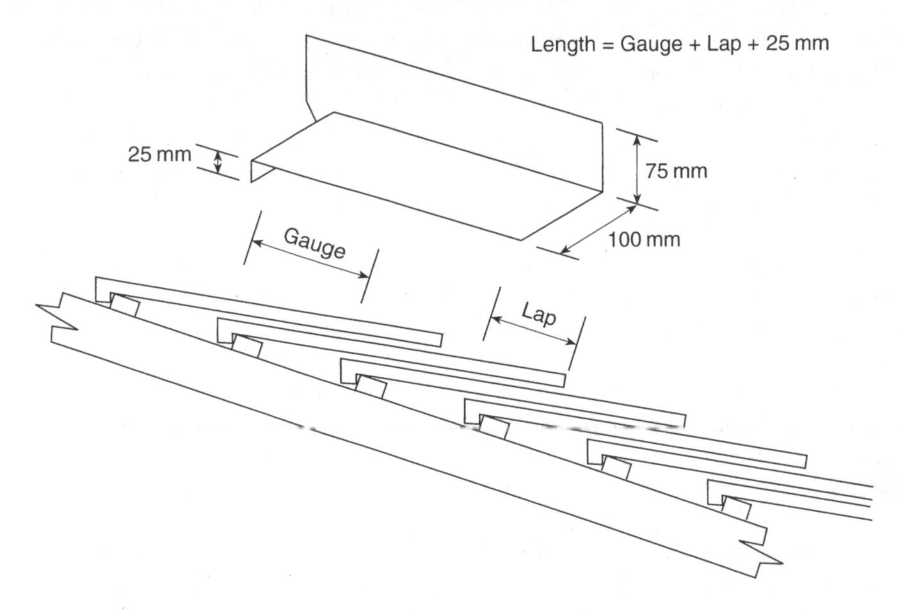

Fig. 13.1 Soaker for tiled roof

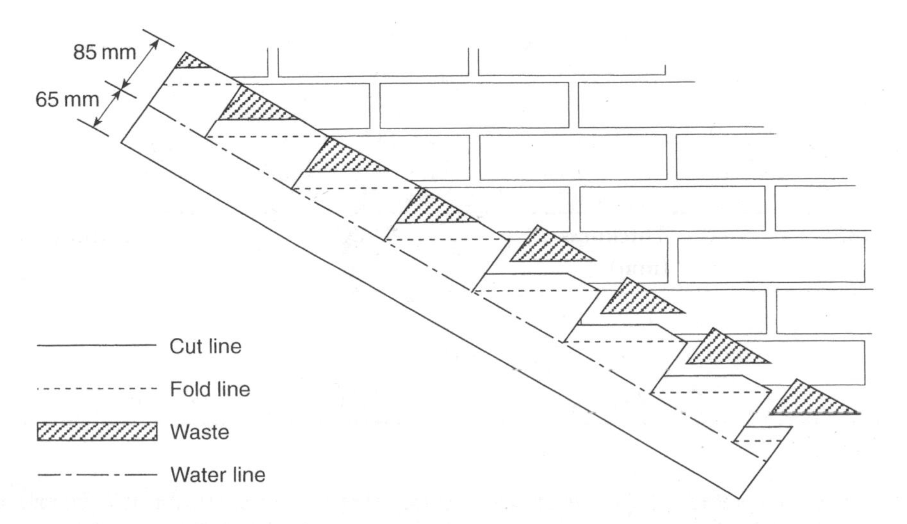

Fig. 13.2 Continuous step flashing

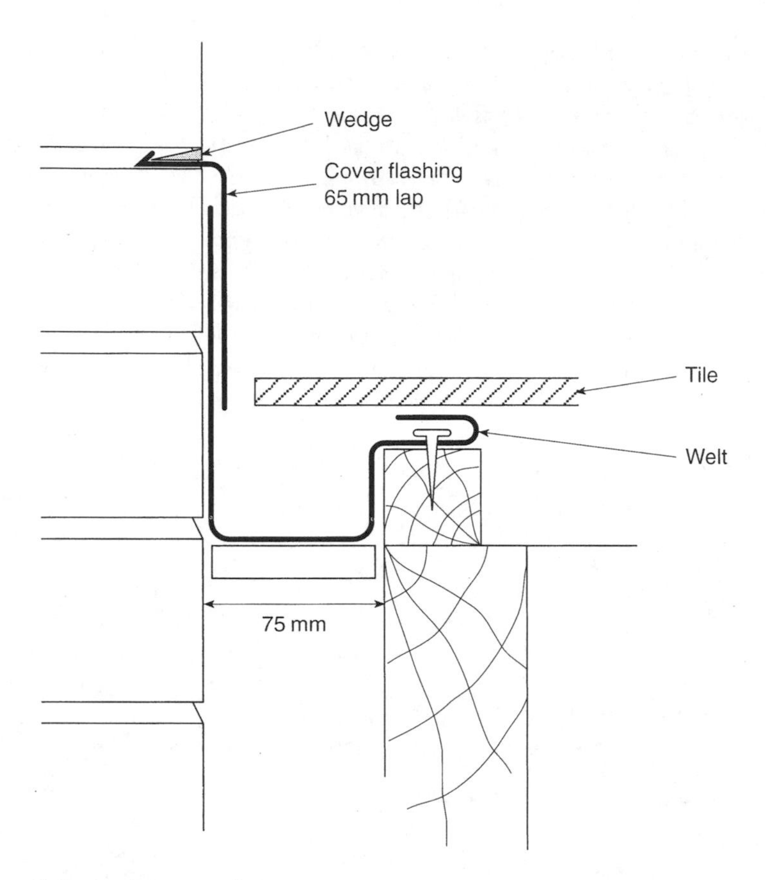

Fig. 13.3 Abutment flashing with secret gutter

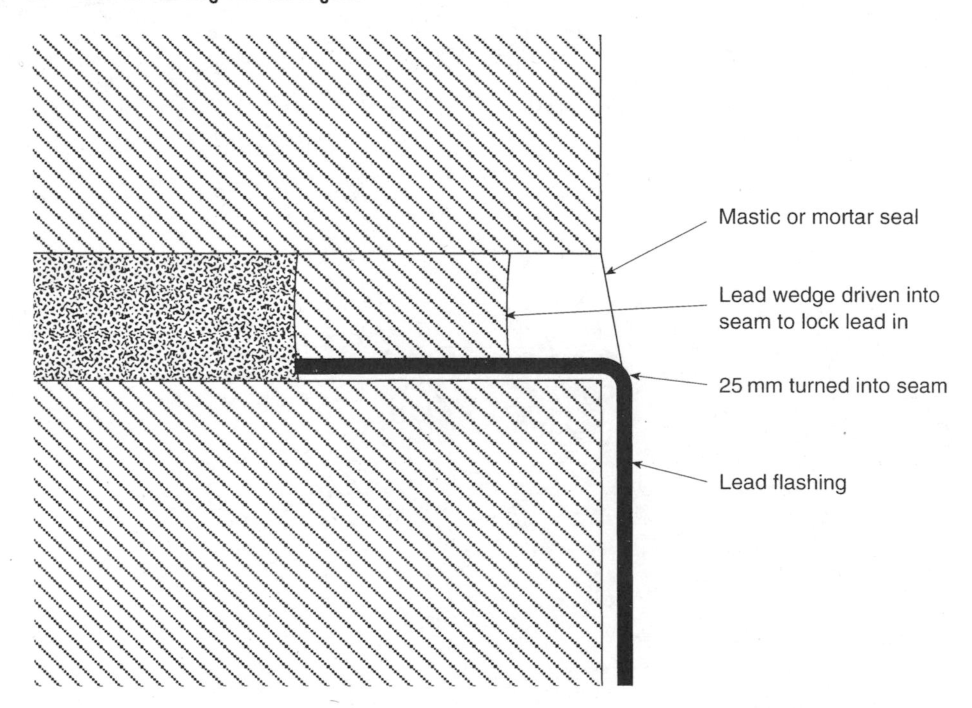

Fig. 13.4 Fixing to masonry

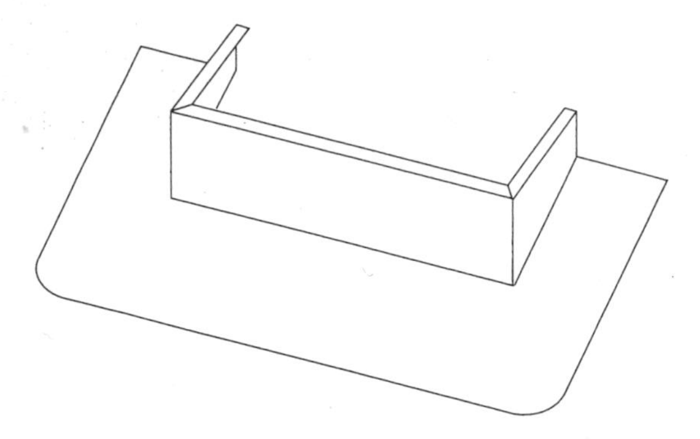

Fig. 13.5 Bossed chimney apron

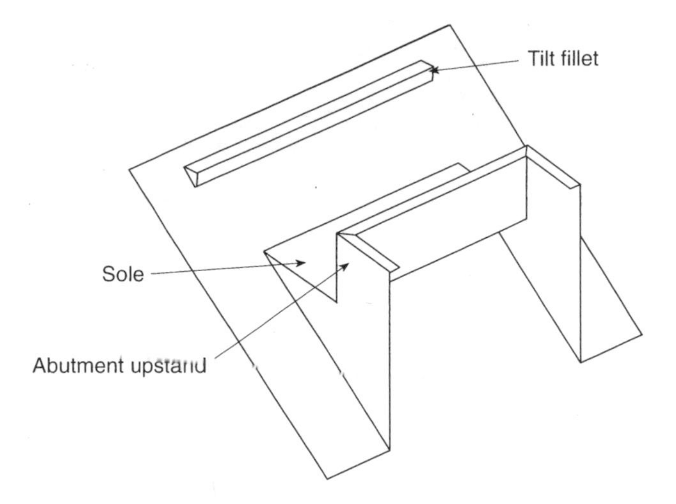

Fig. 13.6 Bossed gutter back

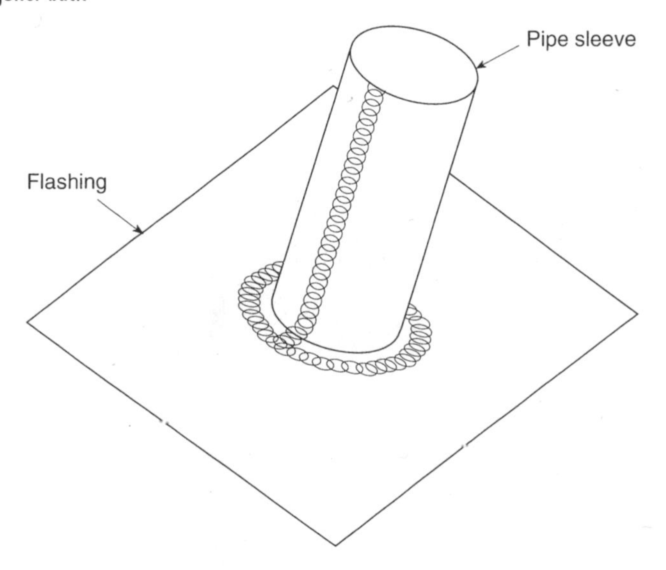

Fig. 13.7 Lead slate with welded joints

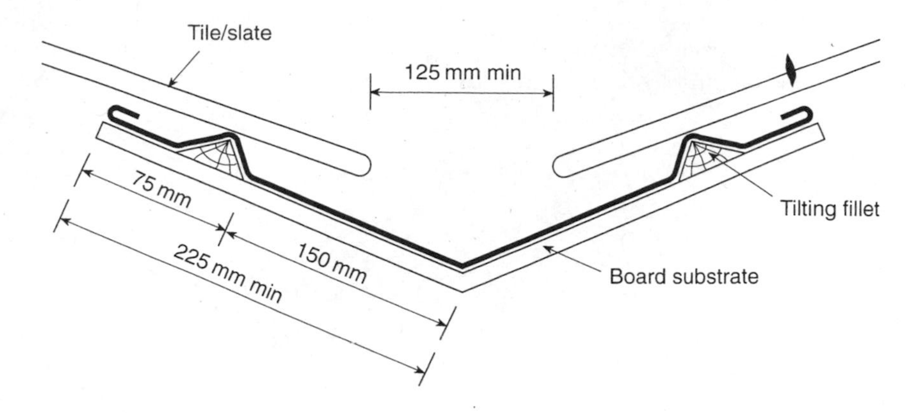

Fig. 13.8 Pitched valley gutter – dimensions

Now try and answer these questions

1. Milled sheet lead for the building process is produced within the criteria set out in
   - **a** BS 1192
   - **b** BS 6644
   - **c** BS 6700
   - **d** BS 1178.

2. The method of identifying the different thicknesses of sheet lead is by a __________ code.
   - **a** stamp
   - **b** colour
   - **c** letter
   - **d** morse.

3. One method of producing sheet lead is to pass a slab of lead many times through
   - **a** an oven
   - **b** a jack hammer
   - **c** a rolling mill
   - **d** a die.

4. The ore from which lead is derived is called
   - **a** galena
   - **b** feldspar
   - **c** duridium
   - **d** baroxide.

5. A metal with high properties of ductility, malleability and resistance to corrosion used in building is called
   - **a** aluminum
   - **b** lead
   - **c** copper
   - **d** zinc.

6. The surface of sheet lead is protected from corrosion by a surface film referred to as
   - **a** lead oxide
   - **b** enamel
   - **c** patina
   - **d** galvanization.

7. The coefficient of expansion of lead is
   - **a** 0.000 011
   - **b** 0.000 005
   - **c** 0.000 025
   - **d** 0.000 029.

8. When Code 4 sheet lead is used as a flashing its maximum length should be
   - **a** 2.0 m
   - **b** 2.5 m
   - **c** 1.5 m
   - **d** 1.0 m

9. When lead is in contact with either copper, stainless steel, iron, zinc or aluminum there will be little risk of failure through

a erosion
b solvency
c electrolytic corrosion
d dilution.

10. In areas prone to lichen growth the failure of sheet lead can be prevented by the application of

a patination oil
b grey paint
c bitumen mastic
d white paint.

11. To prevent chemical corrosion of lead, the lead should never be laid directly on top of

a plastic
b plywood
c building paper
d oak.

12. Patination oil is applied to new lead to prevent the formation of a

a grey lustre
b white carbonate
c red carbonate
d blue lustre.

13. Clips for securing sheet lead flashings manufactured from copper should be no less than

a 50 mm wide and 0.6 mm thick
b 100 mm wide and 2.0 mm thick
c 10 mm wide and 0.1 mm thick
d 15 mm wide and 0.4 mm thick.

14. The minimum recommended length of nails used to secure sheet lead is

a 10 mm
b 19 mm
c 49 mm
d 30 mm

15. When screws are selected to secure sheet lead, they should be manufactured to

a BS 1012
b BS 1192
c BS 1178
d BS 1210.

16. A device used to secure lead flashings to masonry structures is referred to as a

a plug
b screw
c wedge
d clip.

17. The length of a soaker for slate and tile roofs is equal to

a tile length − 25 mm
b slate length + 25 mm
c gauge + lap − 25 mm
d gauge + lap + 25 mm

18. Soakers are normally manufactured from _______ lead.

a Code 3
b Code 8
c Code 6
d Code 5.

19. Cover flashings should have a maximum length of 1.5 m with a lap joint of _________ minimum cover.

a 50 mm
b 75 mm
c 100 mm
d 125 mm.

20. Where an abutment wall is constructed from random built stone the preferred flashing is

a continuous step
b single step
c saddle
d secret gutter.

21. A method of weathering the angle of the joint between two pitched roofs is the used of a lead lined

a valley gutter
b secret gutter
c half-round gutter
d OG gutter.

22. The recommended minimum depth taken out of a seam for fixing flashings is

a 40 mm
b 10 mm
c 35 mm
d 25 mm.

23. Flashings are secured to walls by turning the lead into a seam and hammering in a __________ at 250 mm spacings.

a cleat
b wedge
c nail
d screw.

24. An alternative to using soakers to weather an abutment wall is the use of a

a valley gutter
b secret gutter
c half round gutter
d OG gutter.

25. The type of flashing used to weather the apex of two valley gutters is referred to as a

a apex flashing
b step flashing
c saddle
d bridge.

26. The front of a chimney penetrating a pitched roof is weathered by a flashing called a

a gutter back
b saddle
c bridge
d apron.

27. Traditionally the chimney apron was made using the __________ technique.

a folding
b bossing
c pressing
d drifting.

28. The first flashing to be applied to a chimney is the

a apron
b gutter back
c step
d soaker.

29. The joint between the back of a chimney and the pitch of a roof is weathered by a flashing called a

a box gutter
b saddle
c bridge
d gutter back.

30. Where lead is integrated within the structure of a building to prevent the movement of water, it is referred to as a

a water stop
b moisture trap
c damp proof course
d vapour barrier.

31. The type of flame used for lead welding should be a

a neutral flame
b oxidizing flame
c carbonizing flame
d hot flame.

32. The welding process of lead is of the __________ type.

a brazing
b autogenous
c alloy forming
d oxidizing.

33. The Latin word for lead, from which plumbers get their name is

a galena
b plumbosser

c boraxius
d plumbum.

34. The skill of working sheet lead into complex shapes from flat sheets of lead is termed

a lead dressing
b lead folding
c lead bossing
d lead welding.

35. Where a pipe penetrates a roof the flashing used to weather the penetration is referred to as a

a lead slate
b saddle
c soaker
d pipe sleeve.

36. The lead tool designed for flattening down sheet lead is called a

a mallet
b dresser
c setting-in stick
d bending stick.

37. Traditional lead-working tools are made from

a pitch pine
b mahogany
c oak
d boxwood.

38. A type of joint used on sheet lead which provides a means of allowing expansion to take place is referred to as a

a welded joint
b double lock welt
c lap joint
d bail tack.

39. A lead-working tool which is best suited to moving lead to form corners, etc. is called a

a mallet
b chase wedge
c dresser
d bossing stick.

40. When a piece of lead to be bossed lies on top of another sheet, a ________ should be placed between them.

a plastic sheet
b drift plate
c newspaper
d dinner plate.

41. One method of weathering the ridge of a pitched roof is to have a lead covered

a ridge roll
b ridge tile
c box
d canopy.

42. The alternative to having a lead lined valley gutter is to use ___________ integrated with the slates.

a building paper
b roofing felt
c plastic
d valley soakers.

43. The amount of lap necessary to weather a lap joint is dependent on the

a direction of wind
b angle of slope
c thickness of lead
d type of roof.

44. The flashing used to drain a flat roof with parapet walls is referred to as a

a lead slate
b secret gutter
c chute gutter
d gutter back.

45. When sheet lead is welded it is very important that the surfaces to be welded are prepared to expose

a virgin metal
b the oxide coating
c the colour code
d the welder's skill.

Now check your answers from the grid.

Q 1; d	Q 10; a	Q 19; c	Q 28; a	Q 37; d
Q 2; b	Q 11; d	Q 20; b	Q 29; d	Q 38; c
Q 3; c	Q 12; b	Q 21; a	Q 30; c	Q 39; d
Q 4; a	Q 13; a	Q 22; d	Q 31; a	Q 40; b
Q 5; b	Q 14; b	Q 23; b	Q 32; b	Q 41; a
Q 6; a	Q 15; d	Q 24; b	Q 33; d	Q 42; d
Q 7; d	Q 16; c	Q 25; c	Q 34; c	Q 43; b
Q 8; c	Q 17; d	Q 26; d	Q 35; a	Q 44; c
Q 9; c	Q 18; a	Q 27; b	Q 36; b	Q 45; a

NOTES